AF619059

Automatic Calibration of Modulated Frequency Synthesizers

THE KLUWER INTERNATIONAL SERIES IN ENGINEERING AND COMPUTER SCIENCE

ANALOG CIRCUITS AND SIGNAL PROCESSING

Consulting Editor*: Mohammed Ismail. *Ohio State University

Related Titles:

CONTINUOUS-TIME SIGMA-DELTA MODULATION FOR A/D CONVERSION IN RADIO RECEIVERS
L. Breems, J.H. Huijsing
ISBN: 0-7923-7492-4
DIRECT DIGITAL SYNTHESIZERS: THEORY, DESIGN AND APPLICATIONS
J. Vankka, K. Halonen
ISBN: 0-7923 7366-9
SYSTEMATIC DESIGN FOR OPTIMISATION OF PIPELINED ADCs
J. Goes, J.C. Vital, J. Franca
ISBN: 0-7923-7291-3
OPERATIONAL AMPLIFIERS: Theory and Design
J. Huijsing
ISBN: 0-7923-7284-0
HIGH-PERFORMANCE HARMONIC OSCILLATORS AND BANDGAP REFERENCES
A. van Staveren, C.J.M. Verhoeven, A.H.M. van Roermund
ISBN: 0-7923-7283-2
HIGH SPEED A/D CONVERTERS: Understanding Data Converters Through SPICE
A. Moscovici
ISBN: 0-7923-7276-X
ANALOG TEST SIGNAL GENERATION USING PERIODIC ΣΔ-ENCODED DATA STREAMS
B. Dufort, G.W. Roberts
ISBN: 0-7923-7211-5
HIGH-ACCURACY CMOS SMART TEMPERATURE SENSORS
A. Bakker, J. Huijsing
ISBN: 0-7923-7217-4
DESIGN, SIMULATION AND APPLICATIONS OF INDUCTORS AND TRANSFORMERS FOR Si RF ICs
A.M. Niknejad, R.G. Meyer
ISBN: 0-7923-7986-1
SWITCHED-CURRENT SIGNAL PROCESSING AND A/D CONVERSION CIRCUITS: DESIGN AND IMPLEMENTATION
B.E. Jonsson
ISBN: 0-7923-7871-7
RESEARCH PERSPECTIVES ON DYNAMIC TRANSLINEAR AND LOG-DOMAIN CIRCUITS
W.A. Serdijn, J. Mulder
ISBN: 0-7923-7811-3
CMOS DATA CONVERTERS FOR COMMUNICATIONS
M. Gustavsson, J. Wikner, N. Tan
ISBN: 0-7923-7780-X
DESIGN AND ANALYSIS OF INTEGRATOR-BASED LOG -DOMAIN FILTER CIRCUITS
G.W. Roberts, V. W. Leung
ISBN: 0-7923-8699-X
VISION CHIPS
A. Moini
ISBN: 0-7923-8664-7
COMPACT LOW-VOLTAGE AND HIGH-SPEED CMOS, BiCMOS AND BIPOLAR OPERATIONAL AMPLIFIERS
K-J. de Langen, J. Huijsing
ISBN: 0-7923-8623-X
CONTINUOUS-TIME DELTA-SIGMA MODULATORS FOR HIGH-SPEED A/D CONVERTERS: Theory, Practice and Fundamental Performance Limits
J.A. Cherry, W. M. Snelgrove
ISBN: 0-7923-8625-6
LEARNING ON SILICON: Adaptive VLSI Neural Systems

AUTOMATED CALIBRATION OF MODULATED FREQUENCY SYNTHESIZERS

Dan McMahill
Massachusetts Institute of Technology, U.S.A.

SPRINGER SCIENCE+BUSINESS MEDIA, LLC

ISBN 978-1-4757-8330-8 ISBN 978-0-306-47516-0 (eBook)
DOI 10.1007/978-0-306-47516-0

Library of Congress Cataloging-in-Publication Data

A C.I.P. Catalogue record for this book is available
from the Library of Congress.

Printed on acid-free paper.

Contents

List of Figures

List of Tables

Foreword

Wireless communcation with portable devices is enabled by innovative low power circuit and system techniques implemented in silicon integrated circuits. One of the key ideas to reduce power dissipation is to reduce the number of circuit blocks required for a specific function. This idea is exemplified by the use of modulated frequency synthesizers for digital data transmission rather than conventional quadrature modulated transmitter architectures. Circuit blocks such as multipliers and DACs are not required in this architecture.

The origins of this work come from Miller and Conley at Hewlett Packard and Riley and Copeland at Carlton University. Their work resulted in direct sigma-delta modulation of the feedback divide value in a fractional-N phase locked loop to achieve phase/frequency modulation. This topology maintains the stability of a closed loop system allowing for continuous transmission, while open loop architectures are limited to data bursts on the order of the oscillator drift. One limitation to this architecture is that the loop filter cutoff frequency is constrained to be low to limit noise resulting in rather low data rates.

Perrott's work at MIT broke through this barrier by the use of a digital precompensation filter. With the addition of this filter the modulated signal has an overall flat response well above the loop filter cutoff frequency allowing for higher rate modulation. He demonstrated data rates up to 2.5 Mb/s with this architecture. This architecture was limited by the requirement of matching of the analog loop filter cutoff frequency with the precompensation filter emphasis frequency. Any mismatch resulted in intersymbol interference.

The key contribution of the work presented in this monograph is a technique for in service automatic calibration of the modulated frequency synthesizer by ensuring that the digital emphasis filter and analog loop filter characteristics are matched. The automatic calibration circuit operates while the transmitter is in service and compensates for process and temperature variations. GFSK and 4-GFSK modulation was demonstrated at data rates of 2.5 Mb/s and 5 Mb/s respectively at an RF output carrier frequency of 1.8 GHz.

In this monograph McMahill describes the motivation for constant envelope modulation to eliminate the requirement for linear power amplifiers. He discusses a comparison of a variety of constant envelope modulator architectures along with the key performance requirements for GFSK modulation. He presents an overview of the calibration system and a detailed description of the integrated circuit implementation. He concludes with some impressive experimental results. In addition, he presents some valuable tools for the practicing engineer in this field.

Charles G. Sodini
Professor EECS at MIT

To Miller and Heidi

Acknowledgments

This monograph was originally published as a thesis describing the results of a research project at M.I.T. I would like to thank Carl Harris at Kluwer for encouraging me to publish the work contained in this monograph and John Guttag at M.I.T. for permission to republish my thesis work. Charles G. Sodini, also of M.I.T., kindly wrote the foreword. Some of the material contained in this monograph has been published by the IEEE and I thank them for granting permission to reuse that material.

DARPA provided partial funding for this research project under contract DAAL-01-95-K-3526. Maxim Integrated Products also provided partial support.

Chapter 1

INTRODUCTION

The focus of this monograph is the development of a low power, radio frequency transmitter architecture. Specifically, a technique for in service automatic calibration of a modulated phase locked loop (PLL) frequency synthesizer is presented and examined in detail. Phase/frequency modulation is accomplished by modulating the feedback divide value in a phase locked loop frequency synthesizer [1]. A digital pre-compensation filter is used to extend the modulation bandwidth by canceling the low-pass transfer function of the PLL [2]. The automatic calibration circuit maintains accurate matching between the digital precompensation filter and the analog PLL transfer function across process and temperature variations. The automatic calibration circuit, which is the main contribution of this work, has been designed to operate while the transmitter is in service. This online calibration eliminates the need for production calibration and periodic down time for calibration cycles.

Key features of the new calibration system include the following:

- High data rate Gaussian Frequency Shift Keying (GFSK) and Gaussian Minimum Shift Keying (GMSK) modulation.
- Support for M-ary FSK modulation.
- No required manufacturing calibration.
- Potential for low power operation.

Key features of the implemented system include:

- RF output carrier center frequency in the 1.88 GHz range.
- GMSK/GFSK modulation at 2.5 Mb/s.

- 4-GFSK modulation at 5.0 Mb/s.
- Low power (78 mW) operation.

1.1. Outline

This monograph is organized as follows:

- Motivation for constant envelope modulation
- Comparison of various constant envelope modulator architectures
- Key system performance requirements for GFSK/GMSK
- Overview of the calibration system
- Experimental System Implementation Details
- Test Results
- Conclusions

1.2. Choice of Modulation Affects Battery Life

This section will discuss why constant envelope modulation types are attractive for wireless systems. The type of modulation plays a large role in the efficiency of radio frequency (RF) power amplifiers. The power added efficiency, η, is a key performance parameter of the power amplifier. As given by (1.1), η is defined as the ratio of the added RF power to the signal to the power supplied by the battery [3].

$$\eta = \left(\frac{P_{RF,out} - P_{RF,in}}{P_{DC,in}} \right) \cdot 100\% \tag{1.1}$$

In many wireless communication devices, the power dissipated by the RF power amplifier is a sizeable percentage of the total battery power. When a modulated signal with significant amplitude variations such as quadrature phase shift keying (QPSK) or quadrature amplitude modulation (QAM) is applied to an amplifier which exhibits nonlinearities, two important signal degradations occur. The nonlinear distortion causes distortion to the signal which causes degraded reception by the intended receiver. The second effect is an increase in spurious transmit energy in adjacent channels. This effect is known as spectral regrowth and plays a prime roll in determining the required amplifier linearity. Linearity in power amplifiers has been studied extensively [4] and one of the results is that power added effecienty is sacrificed to obtain higher linearity, see for example [5] and [6]. If the amplitude variations in the modulated signal are reduced, then spectral regrowth is reduced. This reduction in spectral regrowth may be traded for increased operating output power and associated power added efficiency.

The desire for higher efficiency power amplifiers and corresponding increase in battery life has led to the use of modulation types with reduced amplitude variations. Such modulation types include $\frac{\pi}{4}$ differential quadrature phase shift keying ($\frac{\pi}{4}$-DQPSK) used by the IS-54 United States Digital Cellular (USDC) Standard [7], offset quadrature phase shift keying (OQPSK) used by the IS-95 CDMA cellular phone system [7], Gaussian frequency shift keying (GFSK) used by the digital european cordless telephone (DECT) standard [7], and Gaussian minimum shift keying (GMSK) used by the European Global System for Mobile Communications (GSM) system [7]. Of these modulation types, GFSK and GMSK fall in the the continuous phase modulation (CPM) class of constant envelope modulations [8]. It is this class of modulation that can be generated with the modulated synthesizer.

1.3. Classes of Modulation

Before focusing on the particular modulation type of interest for this work, it is useful to examine a few general classes of modulation. Modulation types can be generally classified as either linear modulation or nonlinear modulation.

1.3.1 Linear Modulation

The class of modulation schemes known as pulse amplitude modulation (PAM) are called linear modulation due to the linear relationship between the data symbols and the transmitted signal. When analyzing narrowband signals that arise in communications systems, it is often times convenient to represent the signal as in (1.2).

$$s(t) = \Re\left\{s_{ce}(t)e^{j\omega_c t}\right\} \tag{1.2}$$

The signal $s_{ce}(t)$ is known as the complex envelope of the signal [9] and ω_c is the center frequency of the carrier. For PAM, the complex envelope is given by (1.3) [10].

$$s_{ce}(t) = \sum_k a_k p(t - kT) \tag{1.3}$$

The sequence of (possibly complex) data symbols is given by a_k and $p(t)$ is the (possibly complex) impulse response of a linear time-invariant pulse shaping filter. When the impulse response $p(t)$ is a smooth function, the envelope of the RF output signal, $|s_{ce}(t)|$, varies with time. Examples of linear modulation are QPSK and QAM.

1.3.2 Constant Envelope Modulation

Phase and frequency modulation are both nonlinear modulation types because the output signal does not depend on the modulating signal in a linear fashion. One advantage of this type of modulation is that the envelope can be made inherently constant. The complex envelope for a constant envelope signal is given by (1.4) [10].

$$s_{ce}(t) = e^{j\psi(t)} \tag{1.4}$$

The phase function, $\psi(t)$, is commonly generated by filtering the data sequence, a_k, with a linear time-invariant filter with impulse response, $p(t)$.

$$\psi(t) = \sum_k a_k p(t - kT) \tag{1.5}$$

1.4. GFSK and GMSK

The class of modulations given by (1.4) and (1.5) include two popular and closely related modulation types which are Gaussian frequency shift keying (GFSK) and Gaussian minimum shift keying (GMSK) [11]. For GFSK and GMSK, the pulse shaping filter in (1.5) is given by (1.6),

$$p_{GMSK}(t) = rect_T(t) * g(t) \tag{1.6}$$

where $rect_T(t)$ is a rectangular pulse of duration T, and $g(t)$ is a Gaussian pulse given by (1.7).

$$g(t) = \frac{1}{\sqrt{2\pi\sigma^2}} e^{-\frac{1}{2}\frac{t^2}{\sigma^2}} \tag{1.7}$$

The -3 dB bandwidth of the filter specified by (1.7) is

$$f_{-3dB} = \frac{\sqrt{\log(2)}}{2\pi\sigma} \tag{1.8}$$

and the noise bandwidth is

$$f_N = \frac{1}{4\sigma\sqrt{\pi}} = f_{-3dB}\sqrt{\frac{\pi}{4\log(2)}} \approx 1.065 f_{-3dB}. \tag{1.9}$$

In addition to using the filter specified by (1.6), GMSK and GFSK use a frequency modulation index of $m = 0.5$. The standard definition, see [12] for example, of modulation index for FSK type modulations is given by (1.10).

$$m \triangleq f_{dev,p-p} T_{SYM} \tag{1.10}$$

The symbol period, T_{SYM}, is in seconds and the peak to peak frequency deviation, $f_{dev,p-p}$, is in Hz. Using the definition of modulation index, we can calculate the change in carrier phase during one symbol time from (1.11).

$$\Delta_{\phi} = a_k 2\pi f_{dev,peak} T_{SYM} = a_k \pi m \tag{1.11}$$

The data symbol, a_k, is either 1 or -1. Using a modulation index of $m = 0.5$ causes the phase of the RF carrier to either advance or decrease by $\frac{\pi}{2}$ radians or 90 degrees relative to a carrier center frequency phase reference during one symbol time depending on the value of the transmitted bit. The difference between GMSK and GFSK is that GMSK requires a modulation index of precisely 0.5 while GFSK can work with different modulation indices. However, a nominal modulation index of 0.5 is common for GFSK systems.

GMSK/GFSK is further classified by the product of the Gaussian filter -3 dB bandwidth, B, and the symbol time, T. Values of BT ranging from 0.2 to 0.5 are common in practice. The DECT standard employs GFSK with $BT = 0.5$ [13] and the GSM standard employs GMSK with $BT = 0.3$ [7]. The parameter σ is related to the symbol rate and the parameter BT by

$$\sigma = \frac{\sqrt{\log(2)}}{2\pi f_{SYM} BT}. \tag{1.12}$$

The ratio of noise bandwidth to symbol rate is

$$\frac{f_N}{f_{SYM}} = \left(\sqrt{\frac{\pi}{4\log(2)}}\right) BT. \tag{1.13}$$

The tradeoff in choosing BT is between the level of adjacent channel energy versus the ease of signal detection. In particular, for $BT \geq 0.5$, it is possible to use a simple limiter/discriminator or PLL based frequency demodulator. For applications requiring higher spectral efficiency, lower values of BT may be used but a phase demodulator is required due to the large ISI present in a frequency demodulator output [11].

To illustrate this tradeoff, Figure 1.1 shows the transmit power spectral density for GMSK with $BT = 0.3$ and $BT = 0.5$. The narrower spectrum produced with $BT = 0.3$ is clearly visible. Figure 1.2 shows the instantaneous frequency[1] of the GMSK modulator output for $BT = 0.3$ and $BT = 0.5$. This is the output of an ideal frequency demodulator in

[1] Despite the uncertainty principle problems with "instantaneous frequency" it is still a useful concept. We accept this term as refering to what one puts into an ideal frequency modulator or what one gets out of an ideal frequency demodulator.

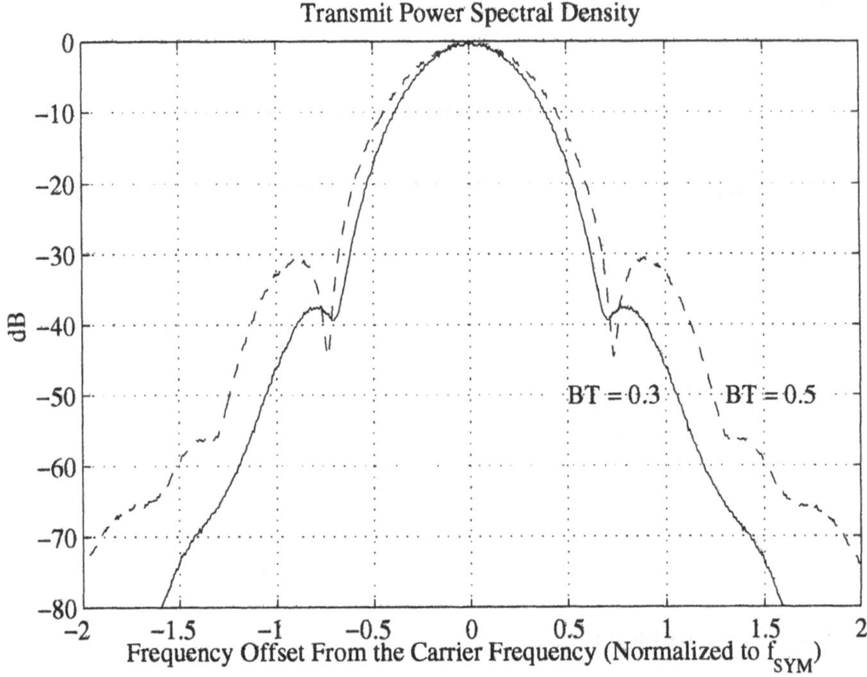

Figure 1.1. GMSK/GFSK Transmit Spectrum

the absence of noise. With $BT = 0.5$, the eye opening is well defined. However, with $BT = 0.3$ the eye is severly closed. This large amount of intersymbol interference (ISI)[14] severly degrades the signal detection. If, however, the received GMSK signal is demodulated with a quadrature demodulator, then the in-phase (I) and quadrature (Q) outputs maintain a large eye opening even with lower values of BT. Figure 1.3 shows the I and Q eye patterns for $BT = 0.3$ and $BT = 0.5$. The detector alternates between looking at I and looking at Q when making symbol decisions.

1.5. Coherent Versus Noncoherent Demodulation

A popular, and general, demodulator architecture for GFSK/GMSK is shown in Figure 1.4. The ideal received signal signal in the absence of noise is given by (1.14).

$$r(t) = \cos\left(\omega_c t + \phi_R(t) + \phi_{mod}(t)\right) \tag{1.14}$$

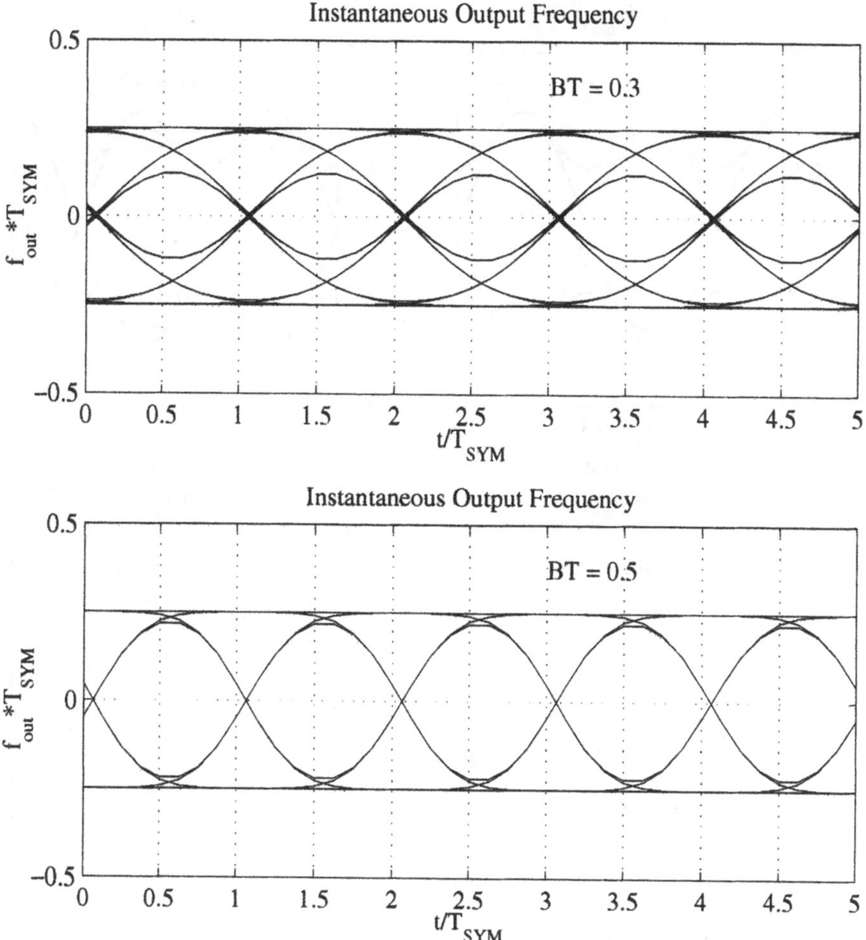

Figure 1.2. GMSK/GFSK Frequency Deviation

The carrier center frequency is given by ω_c, $\phi_R(t)$ is an unknown phase due to uncertainty in the delay through the channel, delay through various circuits, and phase noise of the oscillators, and $\phi_{mod}(t)$ is the modulation part of the phase function. In the case of a coherent demodulator, the DSP section implements a carrier tracking phase locked loop which causes the local oscillator phase to track the phase $\omega_c t + \phi_R(t)$ of the received signal. This arrangement results in the received I and Q waveforms shown in Figure 1.3.

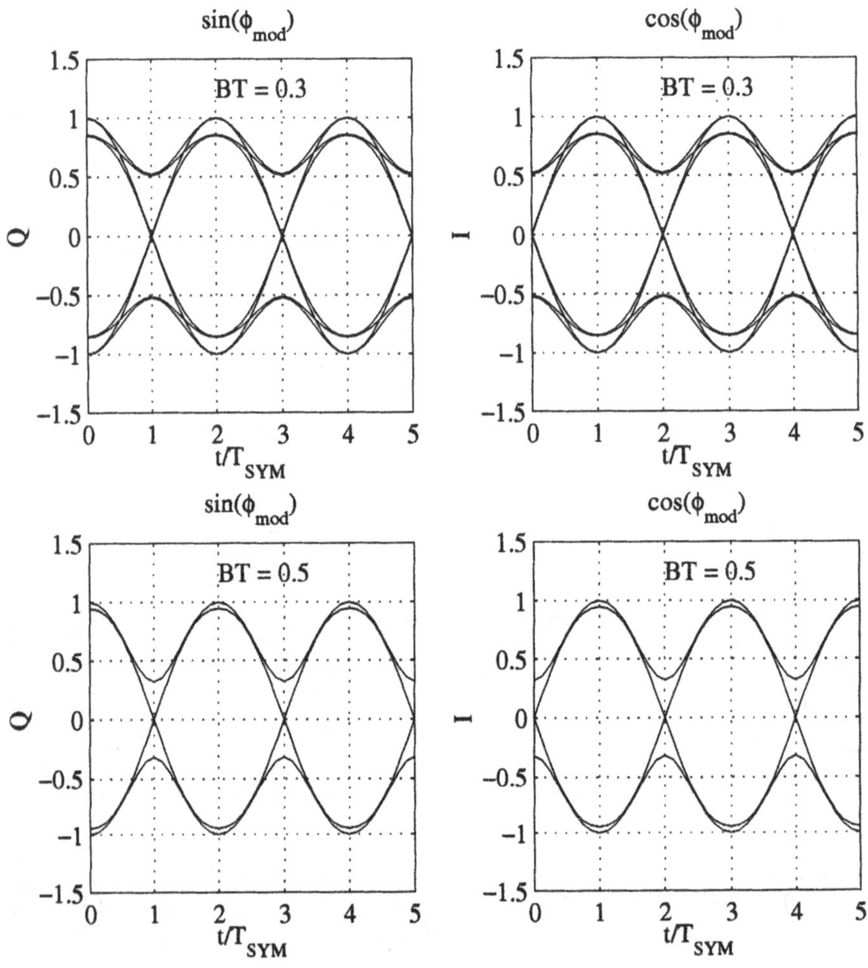

Figure 1.3. GMSK In-phase (I) and Quadrature (Q) Eye Patterns

In a differentially coherent demodulator [14], the local oscillator is simply set to the nominal carrier frequency without any tracking of the received signal phase. The DSP then computes the phase change between adjacent symbols to determine if a mark or space was transmitted. The simplest demodulator is a noncoherent demodulator. In a noncoherent demodulator, the DSP section performs FM demodulation.

Noncoherent demodulation has the advantages of a simpler receiver (no need for a carrier phase detector and other associated PLL components) and fast response for burst transmissions and fading channels

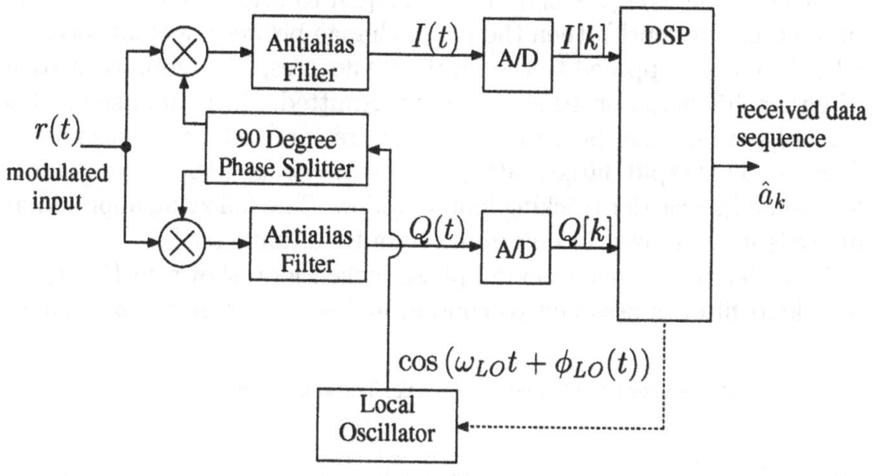

Figure 1.4. Generic Quadrature Demodulator

(there is no need to track the carrier phase). There are, however, disadvantages to noncoherent demodulation. The primary disadvantage is the required input signal to noise ratio required for a given bit error rate performance. The use of coherent detection provides approximately 2 to 3 dB of improvement over noncoherent detection[2]. In a battery operated system, a 3 dB change in the required transmit power can have a significant (up to a factor of 2) effect on battery life. Viewed from another viewpoint, a 2 to 3 dB change in demodulator performance can result in a several order of magnitude change in the bit error rate at the receiver.

1.5.1 Modulator Requirements for Coherent Detection

When used with a coherent demodulator, the modulator must meet several strict requirements. The phase noise in the transmit signal must be sufficiently small to allow accurate tracking by the carrier tracking loop in the receiver. A more stringent, although related, requirement is modulation depth accuracy. Consider a modulator which is ideal in all respects except depth of modulation. The modulation index, m, is

[2]This is true because a coherent demodulator ignores the noise which is orthogonal to the signal where a non-coherent demodulator is effected by all of the noise. The improvement is not precisely 3 dB because the coherent demodulators estimate of carrier phase is not perfect.

assumed to be $0.5 \cdot (1 + \delta_m)$. If δ_m is equal to 0.1, (a ten percent error in modulation depth), then the phase change between symbols becomes ±99 degrees as opposed to the ideal ±90 degrees. Now suppose a string of either 10 marks or 10 spaces is transmitted. In a time span of 10 symbol periods, the phase error in the transmitted signal becomes 90 degrees. This rapid, large scale, phase error change happens too quickly for even a fast carrier tracking loop to follow. The following approximate analysis may be used to determine acceptible limits on δ_m.

Consider the carrier tracking phase locked loop shown in Figure 1.5. The loop filter is assumed to contain at least 1 integrator so that the

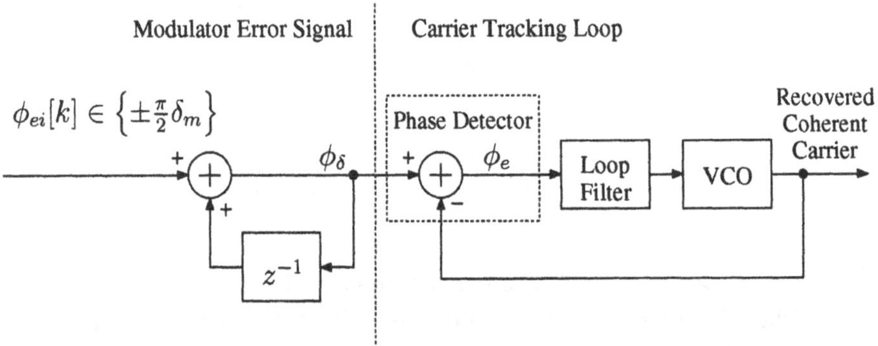

Figure 1.5. Generic Carrier Tracking Phase Locked Loop

static phase error is zero. We can assume that the loop is discrete time in nature operating at a sampling rate which is equal to the symbol rate of the modem. In Figure 1.5, $\phi_{ei}[k]$ represents the sequence of phase change errors in the transmission due to incorrect depth of modulation. The phase of the transmit signal ideally changes by $\pm\frac{\pi}{2}$ radians during each symbol period. When there is a modulation depth error, the phase change differs from the ideal. These symbol by symbol phase change errors are accumulated to produce the total transmitted phase error, ϕ_δ.

Assuming the transmit data sequence is white, ϕ_δ has a power spectral density given by

$$S_{\phi_\delta\phi_\delta}(\theta) = \left(\frac{\pi}{2}\delta_m\right)^2 \frac{1}{2\pi}\left|H_{acc}(e^{j\theta})\right|^2 \tag{1.15}$$

where $H_{acc}(e^{j\theta})$ is the frequency response of the phase accumulator and θ is the normalized angular frequency (ie. $\theta = 2\pi$ corresponds to the sample rate). A few algebraic manipulations give the squared magnitude response of the accumulator given in (1.16).

$$\left|H_{acc}(e^{j\theta})\right|^2 = \left(\frac{1}{2}\right)\left(\frac{1}{1-\cos(\theta)}\right) \tag{1.16}$$

Substituting (1.16) into (1.15) gives the power spectral density of the phase error in the transmit signal. It is this signal that the carrier tracking loop must follow.

$$S_{\phi_\delta\phi_\delta}(\theta) = \left(\frac{\pi}{2}\delta_m\right)^2\left(\frac{1}{2\pi}\right)\left(\frac{1}{2}\right)\left(\frac{1}{1-\cos(\theta)}\right) \tag{1.17}$$

In order to determine the variance of the phase error in the carrier tracking loop, the characteristics of the loop must be known. We can, however, make the following assumption and obtain useful engineering results. If we assume the loop achieves perfect carrier phase tracking within its loop bandwidth and does not track any variations outside of its loop bandwidth, then we can simply integrate (1.17) from $\theta = \theta_c$ to $\theta = (2\pi - \theta_c)$ where θ_c is the normalized loop crossover frequency.

$$\text{var}(\phi_e) \approx \int_{\theta_c}^{2\pi-\theta_c} S_{\phi_\delta\phi_\delta}(\theta)d\theta \tag{1.18}$$

Performing this integration, we obtain (1.19).

$$\text{var}(\phi_e) \approx \left(\frac{\pi}{2}\delta_m\right)^2\left(\frac{1}{2\pi}\right)\left(\frac{1}{\tan(\frac{\theta_c}{2})}\right) \tag{1.19}$$

Assuming that the loop crossover frequency is much less than the sample rate and recognizing that $\tan(x) \approx x$ for $|x| \ll 1$ we can simplify (1.19) and get (1.20).

$$\text{var}(\phi_e) \approx \left(\frac{\pi}{2}\delta_m\right)^2\left(\frac{1}{2\pi}\right)\left(\frac{1}{\frac{\theta_c}{2}}\right) \tag{1.20}$$

This result can be further parameterized by defining γ for the loop to be the ratio of the symbol rate to loop bandwidth. Substituting $\theta_c = \frac{2\pi}{\gamma}$ into (1.20) gives the final result in (1.21).

$$\phi_{e,RMS} \approx \delta_m\sqrt{\frac{\gamma}{8}} \tag{1.21}$$

If an agressive carrier tracking loop bandwidth of approximately 1% of the symbol rate is used (ie. γ is 100), then (1.21) would indicate that δ_m must be less than approximately 0.05 or 5% to achive a RMS phase tracking error of less than approximately 10 degrees. A simulation of an idealized second order carrier tracking loop with a phase margin of 60 degrees and loop crossover frequency of 0.01 times the symbol rate shows a limit of 3% on δ_m for 10 degrees of RMS phase error at the receiver which validates the aproximate result in (1.21). This is a fairly large RMS phase error and it would consume an entire error budget. Allowed levels of phase error will be discussed in a later section. The DECT standard [15] specifies an accuracy in modulation depth of +40%/-10% which indicates that DECT compliancy alone is not adequate for use with coherent detection.

1.6. Chapter 1 Summary

This chapter has provided a motivation for the use of constant envelope modulation. Constant envelope modulation allows the use of more power efficient RF power amplifiers. A common type of constant envelope modulation, GFSK/GMSK, has been discussed in detail.

Chapter 2

ARCHITECTURES

This chapter will examine several possible architectures for generating a GFSK/GMSK signal. All of the architectures that will be discussed in this section are forms of direct modulation. In this context, direct modulation means that the output frequency of the modulator is at the desired RF carrier frequency and no final frequency conversion is required. The reason for omitting superheterodyne architectures is that the power consumption of the frequency converter is moderate or even large compared to that of the modulator and hence undesirable from a low power perspective.

2.1. Direct VCO Modulation

The simplest approach for implementing a GFSK/GMSK modulator is to simply pass the data sequence, a_k, through a zero order hold and a Gaussian pulse shape filter and apply the filtered signal to the input of a voltage controlled oscillator. This approach is illustrated in Figure 2.1. Despite the simplicity of this approach, it is not used in practice due to two major performance limitations. The primary limitation is that of

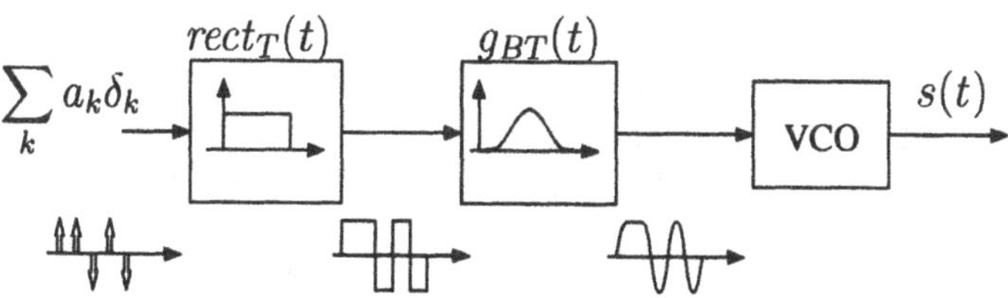

Figure 2.1. Generic Frequency Modulator

center frequency stability. Consider an example system with a carrier center frequency of 1.8 GHz and a channel bandwidth of 3.6 MHz. As a percentage of the center frequency, the channel bandwidth is 0.2% or 2000 ppm. In order to achieve a center frequency error which is on the order of 1% of the channel bandwidth, an oscillator with a center frequency accuracy and stability of on the order of 20 ppm or less is required. With todays technology, this level of accuracy and stability requires the use of a crystal based frequency reference. This requirement is not compatible with the tuning range requirement or programmability requirement.

In addition to the difficulty in achieving the required level of frequency accuracy and stability, the modulation index is set by the characteristics of the D/A converter and by the slope of the VCO tuning curve. As discussed in section 1.5.1, GMSK places stringent requirements on the modulation index. However, it is difficult to control the VCO gain to the required accuracy and stability.

2.2. Quadrature Modulator

The most general architecture is the quadrature modulator shown in Figure 2.2. This approach has enjoyed commercial success and several semiconductor manufacturers offer integrated circuits performing the quadrature modulation [16, 17, 18]. In addition to the commercially available integrated circuits which contain the two RF mixers, phase splitter, and combiner, there exist implementations of the baseband filtering, phase accumulation, sin/cos lookup and the DAC's on a single IC [19, 20]. In the quadrature modulator approach, the desired phase function, $\psi(t)$ is generated digitally. The output of the quadrature modulator as a function of time is then given by

$$s(t) = \cos\left(\omega_c t + \psi(t)\right). \tag{2.1}$$

The local oscillator is generated by a phase locked loop based frequency synthesizer in practical RF systems. The use of the PLL synthesizer is required to achieve the necessary center frequency accuracy while maintaining the ability to change the output frequency. While very general in nature, the quadrature modulator approach has several notable drawbacks. The two RF mixers have stringent requirements on linearity to control adjacent channel interference. The gain and phase through the in phase (I) and quadrature (Q) signal paths must be tightly matched. Mismatch between the I and Q paths as well as phase errors in the 90 degree phase splitter will cause the envelope of the RF signal to experience fluctuations. When passed through a nonlinear amplifier, these envelope fluctuations will cause spectral spreading. In addition to the

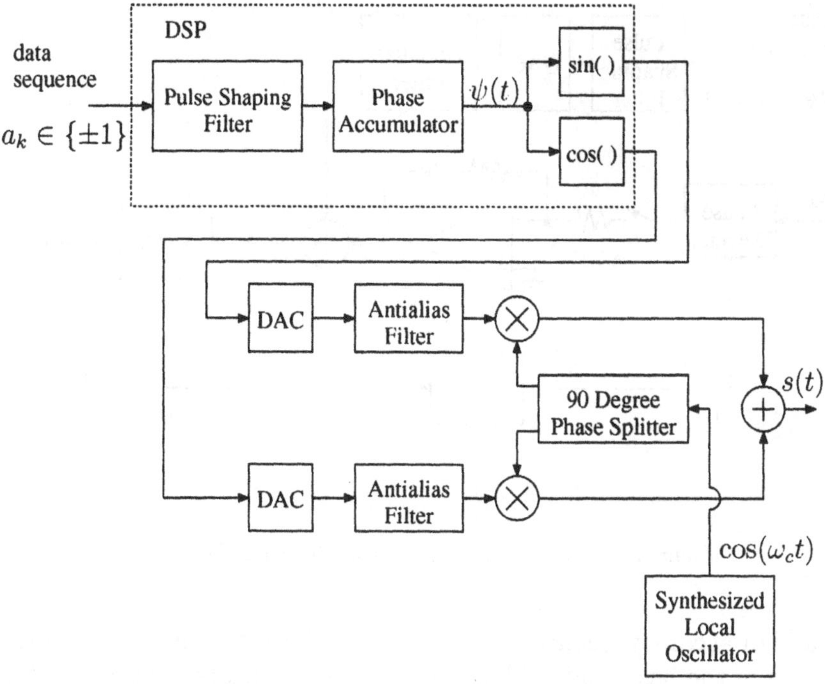

Figure 2.2. Quadrature Modulator

spectral spreading, the phase function becomes distorted. These effects are analyzed in [21]. To meet the required level of performance in the noted areas, the RF mixers and quadrature network consume moderate amounts of power. An additional drawback to the quadrature modulator approach is the requirement for two digital to analog converters and associated antialiasing filters. These two signal paths must have identical characteristics and high dynamic range.

2.3. Open Loop Modulation with PLL Tuning

Figure 2.3 shows another approach which is employed commercially [22] for GFSK systems. In this system, the modulation signal is disabled and the switch closed for a period of time which is sufficient for the PLL to lock and settle. This pretunes the VCO to the correct center frequency of Nf_{ref}. After the PLL has settled, the switch is opened and a data sequence is applied. During modulation, the nominal VCO tuning voltage is held by the capacitor in the loop filter. After the data has been transmitted, the switch is closed to retune the VCO. This approach has several attractive features. The need for high performance RF mixers

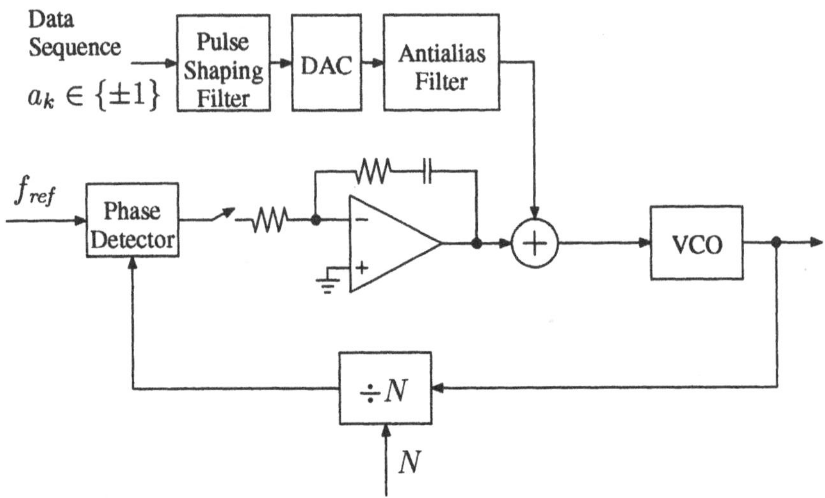

Figure 2.3. Open Loop Modulator with PLL Tuning

and a high accuracy quadrature network has been eliminated. In addition, one of the baseband D/A converters and associated antialias filter has been removed. The output signal is inherently continuous phase and constant envelope.

Despite the attractive features of the system in Figure 2.3, there are several key performance disadvantages. This architecture is not capable of a continuous transmission. The PLL must be periodically allowed to operate to keep the VCO center frequency from drifting outside of its allowed range. During modulation, the VCO is running open loop which makes it more susceptible to disturbances (power supply, mechanical, etc.). In addition, the modulation index is set by the VCO gain and the D/A output characteristics. Component tolerances and drift prevent the achievement of the level of accuracy required by GMSK when coherent detection is desired. As such, this approach is limited to GFSK systems.

2.4. Modulated Fractional–N Synthesizer

A very attractive architecture is based on direct modulation of a fractional–N frequency synthesizer. The use of Σ–Δ modulation for fractional–N frequency synthesis was proposed by [23]. The Σ–Δ fractional-N synthesizer was used as a modulator by [1] and later extended by [24]. At this time the known commercial implementations [25] have modulation bandwidths limited by the PLL loop bandwidth.

The two key insights that led to the development of this architecture are the following. The PLL frequency synthesizer show in Figure 2.4

produces an output frequency equal to N times a reference frequency. This may be viewed as a digital to frequency converter with a coarse

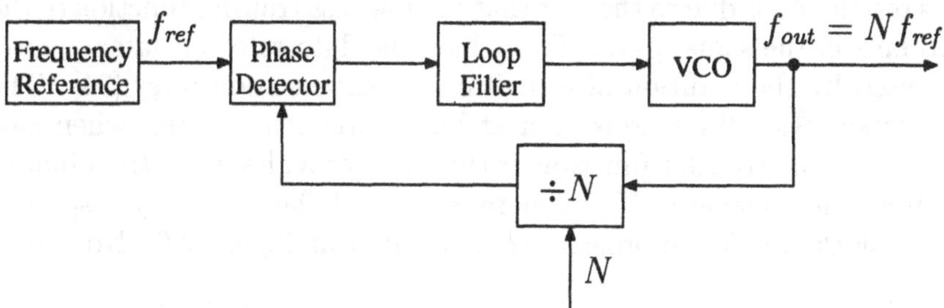

Figure 2.4. Generic PLL Based Synthesizer

step size. The second insight is that noise shaping techniques found in oversampled data converters may be applied to produce a high resolution digital to frequency converter. By simply applying a modulating signal to the input of the frequency to digital converter, very accurate frequency modulation may be achieved. The resulting modulated fractional-N synthesizer is shown in Figure 2.5.

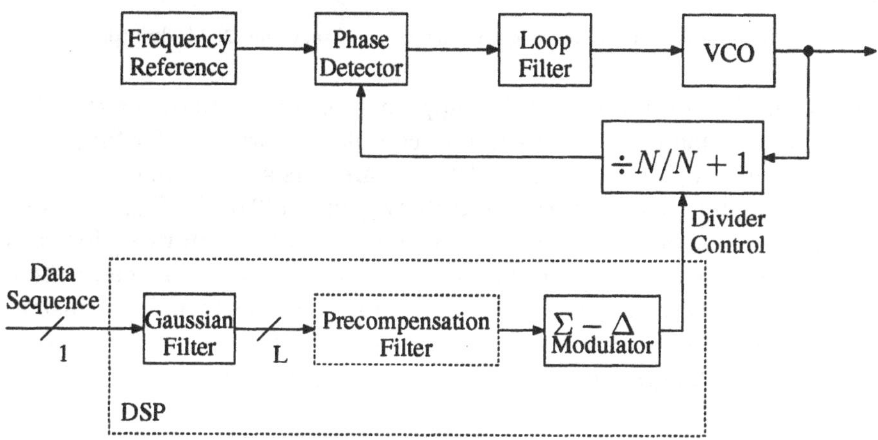

Figure 2.5. Modulated Fractional-N Synthesizer

In the modulated synthesizer approach shown in Figure 2.5, the value of the frequency divider (either N or $N+1$) is controlled by the output of a digital Σ–Δ modulator. The average value of the divider then takes on a non-integer value somewhere between N and $N+1$. The quantization noise from the Σ–Δ modulation is filtered by the lowpass transfer function of the PLL. The combination of the $\Sigma - \Delta$ modulator and PLL forms an accurate digital to frequency converter. In the original paper [1], the data rate was restricted by the loop bandwidth of the PLL. This restriction is due to the fact that the lowpass transfer function of the PLL falls in the signal path. The achievable data rate was later greatly increased by the addition of a digital precompensation filter [24]. The precompensation filter adds gain at higher frequencies and, when cascaded with the transfer function of the PLL, provides a net (nominally) flat frequency response. The general shapes of the frequency responses of the blocks in this approach are illustrated in Figure 2.6. Note that

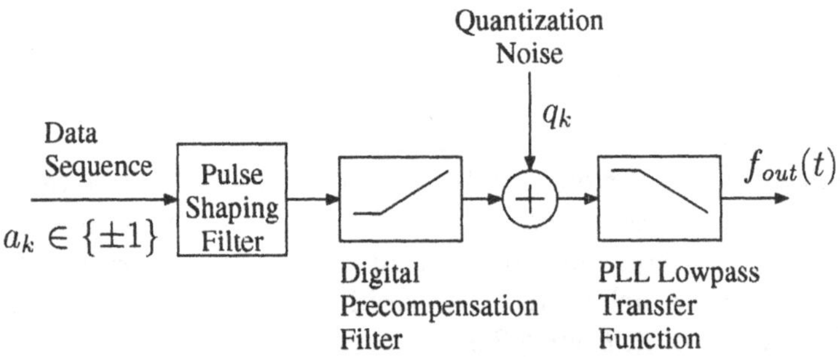

Figure 2.6. Simplified Signal Path of the Fractional-N Modulator

one of the key drawbacks of this approach is the requirement that the precompensation filter and PLL have complementary transfer functions. This requires tight control of the PLL parameters such as phase detector gain, loop filter characteristics, and VCO gain. In [24], the PLL included a digitally controlled gain stage. The gain was then calibrated for each RF channel and the required digital control words stored in a calibration PROM. However, this calibration procedure requires significant instrumentation and is strictly a manufacturing calibration. Any temperature variations in the system must be small to keep the calibration constant correct.

2.5. Calibrated Synthesizer Architecture

The purpose of this section is to outline the new transmitter architecture. This research was based on the extended data rate fractional-N synthesizer in [24] which was briefly discussed in section 2.4. Figure 2.7 shows the basic block diagram of the new architecture. The goal of the research was to design an adaptive tuning mechanism for fine tuning the matching between the digital precompensation filter and the PLL transfer function. The tuning circuitry operates while the modulator is in service. The goal is to eliminate the required factory calibration of each IC and to provide enhanced accuracy in the adjustment. The adaptive

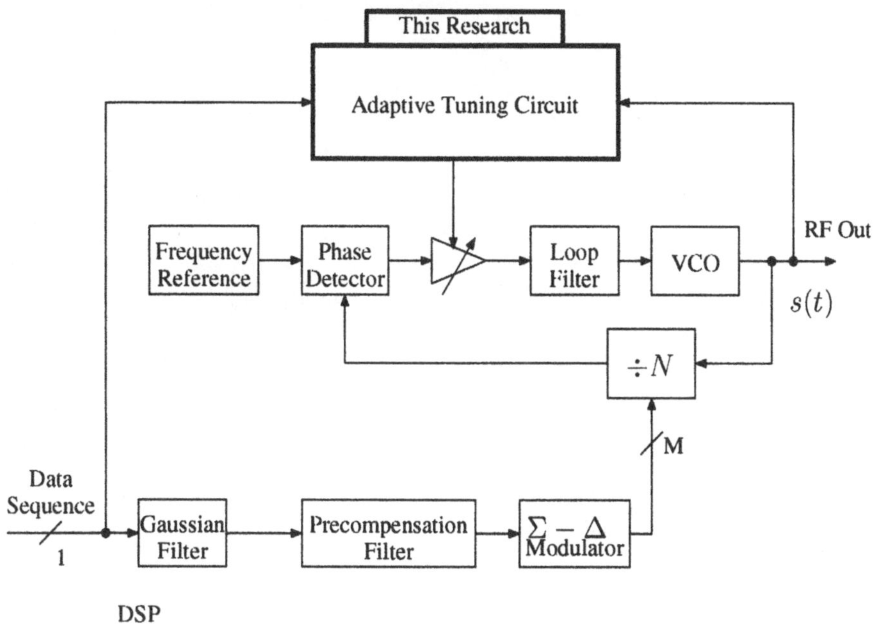

Figure 2.7. Proposed Architecture

tuning circuit, or autocal circuit, operates by monitoring the RF output signal and adjusting the amplitude of the phase detector output current. We will see in a later section that this single adjustment is sufficient to obtain the desired level of matching.

2.6. Chapter 2 Summary

This chapter has reviewed various modulator architectures which may be used for GFSK/GMSK modulation. While offering high performance, the quadrature modulator is relatively power hungry. The open loop modulator consumes less power than the quadrature modulator but has some performance limitations. The Σ–Δmodulated synthesizer offers high performance and low power but some type of calibration circuit is required.

tuning circuit. A [illegible] circuit operates by monitoring [illegible] signal and adjusting the amplitude of the phase detector output [illegible]. We will see in a later section that this [illegible] adjustment is sufficient to obtain the desired level of matching.

2.6 Chapter 2 Summary

This chapter has reviewed [illegible] which may be [illegible] SK/GMSK [illegible] performance, the [illegible] modulator [illegible] lower [illegible] open loop [illegible] type 2 [illegible] required.

Chapter 3

SYSTEM REQUIREMENTS

This chapter will highlight the major system requirements for the modulator.

3.1. Total Phase Error Variance

The primary specification which has been addressed by this research is the phase error in the transmitted signal. The transmitted signal, $s(t)$, can be written as a sinusoidal signal whose phase is made up of the desired phase function and an undesired phase error term

$$s(t) = \cos\left(\omega_c t + \phi_{mod}(t) + \phi_{TX}(t)\right) \tag{3.1}$$

where $\phi_{TX}(t)$, represents any deviation from the ideal phase, ω_c is the carrier center frequency, and $\phi_{mod}(t)$ is the modulation part of the phase.

There are several causes for a deviation from the ideal desired phase trajectory in the transmit signal. Phase noise in any oscillators used in the transmitter adds a random phase to the desired phase. An error in modulation index will cause a random walk in the phase. In addition any deviation between the true frequency response of the precompensated PLL and an ideal flat response will cause errors. The non-ideal frequency response is of primary concern with the chosen architecture due to its large influence on the error in the phase trajectory.

The signal at the demodulator input, $r(t)$, is given by

$$r(t) = \cos\left(\omega_c t + \phi_{mod}(t) + \phi_{RX}(t)\right) + n(t). \tag{3.2}$$

The received signal may contain additional phase errors due to phase noise in any frequency conversions which may have been applied to the signal. The total received phase error signal is $\phi_{RX}(t)$. In addition to any added phase noise, the channel will add a noise signal, $n(t)$. For the purposes of this research, $n(t)$ is assumed to have a Gaussian distribution and have a power spectral density which is flat over the channel bandwidth. In a quadrature demodulator, the received signal, $r(t)$ is multiplied by an in phase (I) local oscillator signal, $LO_I(t)$, and a quadrature (Q) local oscillator signal, $LO_Q(t)$. The two oscillator signals are given by (3.3) and (3.4).

$$LO_I(t) = \cos(\omega_c t + \phi_{LO}(t)) \tag{3.3}$$

$$LO_Q(t) = -\sin(\omega_c t + \phi_{LO}(t)) \tag{3.4}$$

The demodulated I and Q signals are given by (3.5) and (3.6) after the $2\omega_c$ components at the mixer outputs have been removed with a lowpass filter.

$$I(t) = 2r(t)LO_I(t) = \cos(\phi_{mod}(t) + \phi_{RX}(t) - \phi_{LO}(t)) + n_I(t) \tag{3.5}$$

$$Q(t) = 2r(t)LO_Q(t) = \sin(\phi_{mod}(t) + \phi_{RX}(t) - \phi_{LO}(t)) + n_Q(t) \tag{3.6}$$

Under the assumption that the noise at the input to the receiver is Gaussian and white across the channel bandwidth, the baseband noise signals, $n_I(t)$ and $n_Q(t)$ will be independent, white, Gaussian noise signals across the baseband signal bandwidth.

In an ideal coherent receiver, the local oscillator phase, $\phi_{LO}(t)$, exactly tracks $\phi_{RX}(t)$. Due to limited bandwidth in the carrier tracking loop and phase noise in the local oscillator, the tracking is not perfect. We can define a total residual phase error, ϕ_e as given by (3.7).

$$\phi_e(t) \triangleq \phi_{RX}(t) - \phi_{LO}(t) \tag{3.7}$$

In order to assess the effects of ϕ_e on the bit error performance of the system, we first examine the received signal in the absence of a phase error. The ideal I and Q signals, $I_{nom}(t)$, and $Q_{nom}(t)$, are given by (3.8) and (3.9).

$$I_{nom}(t) = \cos(\phi_{mod}(t)) \tag{3.8}$$

$$Q_{nom}(t) = \sin(\phi_{mod}(t)) \tag{3.9}$$

Applying some trigonometric identities to (3.5) and (3.6), gives the more useful form given in (3.10) and (3.11) for the I and Q signals in the presence of a phase error.

$$I(t) = I_{nom}(t)\cos(\phi_e(t)) - Q_{nom}(t)\sin(\phi_e(t)) + n_I(t) \tag{3.10}$$

$$Q(t) = I_{nom}(t)\sin(\phi_e(t)) + Q_{nom}(t)\cos(\phi_e(t)) + n_Q(t) \tag{3.11}$$

Figure 3.1 shows the equivalent baseband system showing the effects of ϕ_e, the pulse shaping filtering, and noise. The $BT = 0.63$ Gaussian

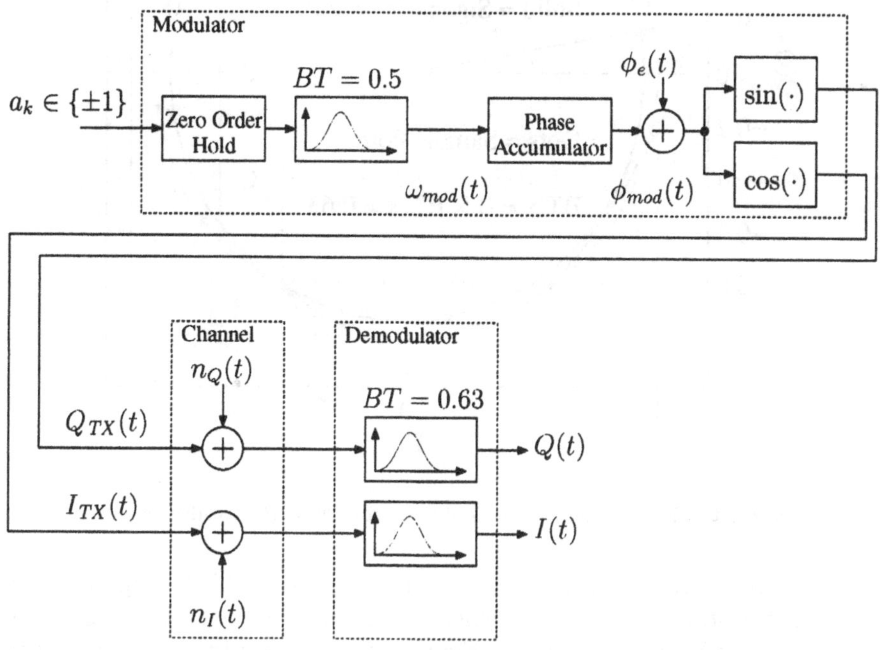

Figure 3.1. System for Bit Error Probability Analysis

filters in the receiver limit the noise bandwidth of the receiver. In [11] it was shown that $BT = 0.63$ in the receiver provides a good compromise between minimizing the noise bandwidth and also minimizing ISI.

A useful way to view the received signal is to plot I(t) versus Q(t). Figure 3.2 shows the received signal constellation in the absence of noise. The signal at the optimal sample time is highlighted in the plot. In Figure 3.2, the residual phase error, ϕ_e is zero. We see from (3.10) and (3.11) that a phase error simply rotates the received constellation plotted in Figure 3.2.

The calculation of bit error probability in the absence of any error control coding is a straightforward task. Due to the symmetry of the

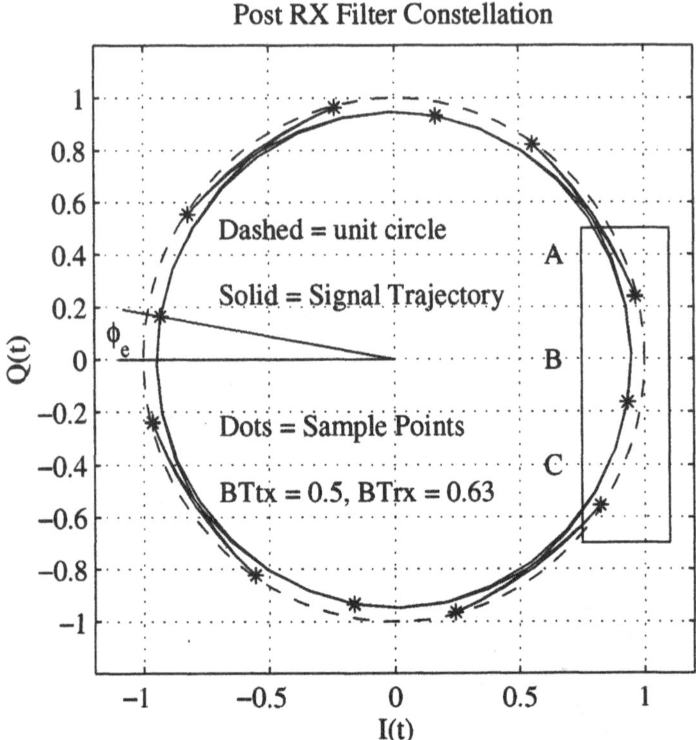

Figure 3.2. Receive Signal Constellation for Error Probability Analysis

signal constellation, the total bit error probability will be equal to the conditional probability of error conditioned on the ideal signal point being one of the three in the box in Figure 3.2. This conditional probability of error can be expanded as

$$P_E = P_{E|\square} = P_{E|A}P_{A|\square} + P_{E|A}P_{B|\square} + P_{E|A}P_{C|\square}. \tag{3.12}$$

The total probability of an error is P_E and the conditional probability of an error given that one of the 3 points in the box is the true signal point is $P_{E|\square}$. The conditional probabilities of each of the 3 signal points are $P_{A|\square} = P_{C|\square} = \frac{1}{4}$ and $P_{B|\square} = \frac{1}{2}$. The evaluation of $P_{E|A}$ can be done by modifying the expression for bit error probability for binary antipodal signaling [8, 14].

$$P_{E,antipodal} = Q\left(\sqrt{\frac{2E_b}{N_o}}\right) \tag{3.13}$$

The energy per bit, E_b, is equal to the signal power times the symbol period. The noise power spectral density at the receiver input is N_o.

The ratio $\frac{E_b}{N_o}$ is a commonly used signal to noise ratio for digital communications systems. The function $Q(x)$ is the probability of a zero mean, unit variance Gaussian random variable exceeding x. By adding two correction factors to (3.13) we obtain the following expression for $P_{E|A}$.

$$P_{E|A} = \int_{-\infty}^{+\infty} P_{E|A,\phi_e} f_{\phi_e}(\theta) d\theta \tag{3.14}$$

The probability density function for ϕ_e is f_{ϕ_e}.

$$P_{E|A,\phi_e} = Q\left(\sqrt{\frac{\gamma_s(\phi_e)\gamma_n 2E_b}{N_o}}\right) \tag{3.15}$$

The factor $\gamma_s(\phi_e)$ accounts for the reduction in the signal power at the sample time from unity. Numerically, $\gamma_s(\phi_e)$ is equal to the square of the projection of point A onto the I-axis. The factor γ_n accounts for the larger noise bandwidth of the Gaussian receive filter compared to the ideal Nyquist bandwidth. Some algebraic manipulations show that for a Gaussian filter with a given BT product, the noise bandwidth is given by (3.16).

$$f_N = \sqrt{\frac{\pi}{4\log_e(2)}} BT f_{SYM} \tag{3.16}$$

The symbol rate, f_{SYM}, is given in symbols per second and the noise bandwidth, f_N, is in Hz. For $BT = 0.63$, (3.16) gives $f_N = 0.671 f_{SYM}$. The ideal noise bandwidth is $\frac{f_{SYM}}{2}$ and hence $\gamma_n = \frac{0.5}{0.671} = 0.746$ for the system shown in Figure 3.1. The choice of $BT = 0.63$ in the receiver is based on the recommendation given in [11] and represents a trade off between increased noise for large BT and increased intersymbol interference for small BT. The two remaining conditional error probabilities, $P_{E|B}$ and $P_{E|C}$, are computed in a similar fashion.

Evaluating (3.12) numerically for several constant values of ϕ_e, corresponding to DC offsets in the receiver carrier tracking loop, yields the bit error probabilities shown in Figure 3.3. The lower plot in Figure 3.3 shows the extra input signal to noise required to achieve a certain bit error probability in the face of a static phase error. Figure 3.4 plots the additional signal to noise ratio required for error probabilities of 10^{-3}, 10^{-4}, and 10^{-5} versus the static phase error. If we take ϕ_e to be a zero mean Gaussian random variable and evaluate (3.12) numerically for several RMS values of ϕ_e, the bit error probabilities in Figure 3.5 result. As in the static phase error case, Figures 3.5 and 3.6 show the additional signal to noise ratio required to achieve a given bit error probability. Figure 3.6 indicates that a total residual phase error on the order of 10

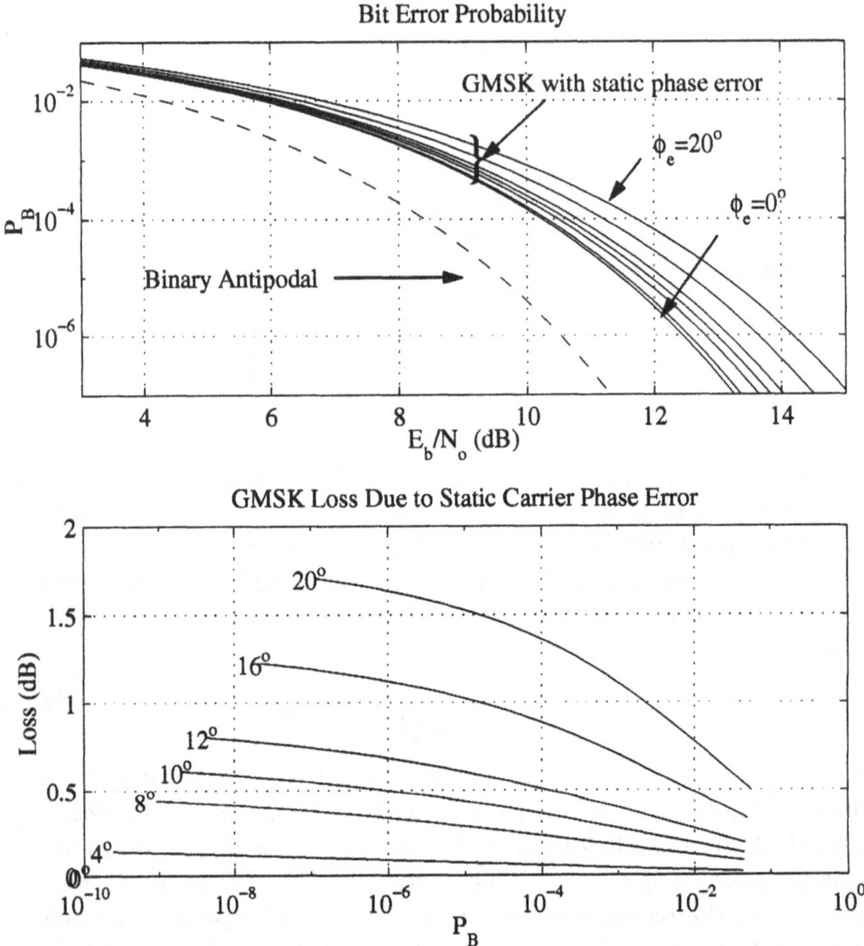

Figure 3.3. Bit Error Probability With 0°, 4°, 8°, 10°, 12°, 16° and 20° Static Phase Error

degrees will degrade the performance of the modem by less than 1 dB for error probabilities of 10^{-5}.

At this point it is worth examining some of the key assumptions made by this analysis. Of primary interest is the modeling of the residual phase error as a Gaussian distributed random variable. There are several contributors to ϕ_e. Phase noise in the local oscillators is due to thermal and shot noise and should be largely Gaussian in nature. However, the phase noise due to mismatch between the digital precompensation filter and the PLL transfer function is not Gaussian. This component of ϕ_e

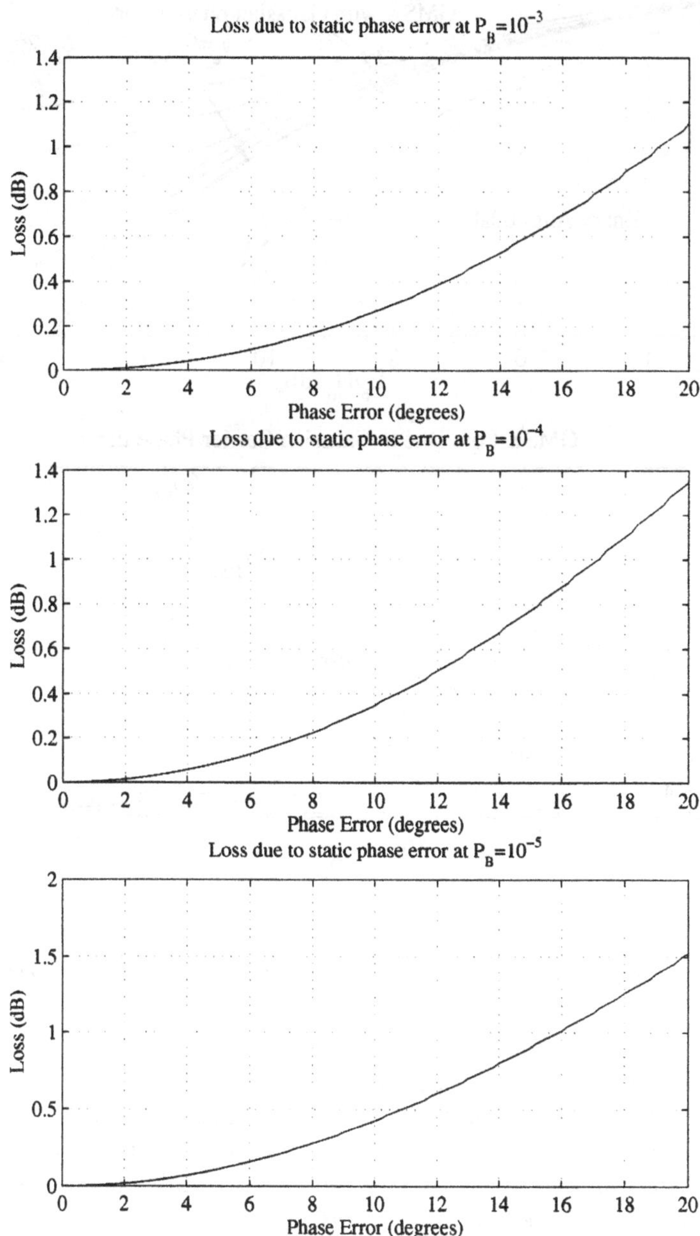

Figure 3.4. Signal to Noise Loss Due to Static Phase Error

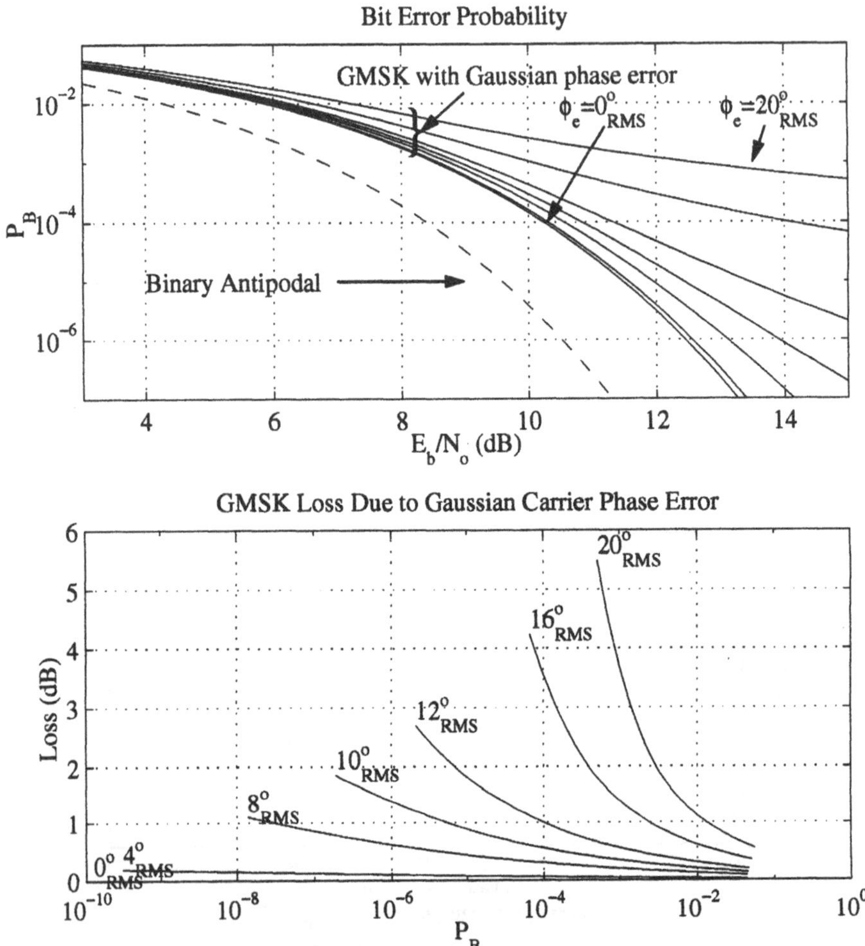

Figure 3.5. Bit Probability Rate With 0°, 4°, 8°, 10°, 12°, 16° and 20° RMS Gaussian Phase Error

is the result of a random telegraph wave (i.e. a ±1 valued waveform which can only transition at the edges of each symbol period) which has been passed through a LTI filter. The LTI filter in effect adds several independent random variables and the central limit theorem suggests that the result may look approximately Gaussian.

3.2. Chapter 3 Summary

This chapter has examined the effects of phase error, both static and random, on the bit error probability of a GMSK system. The result of

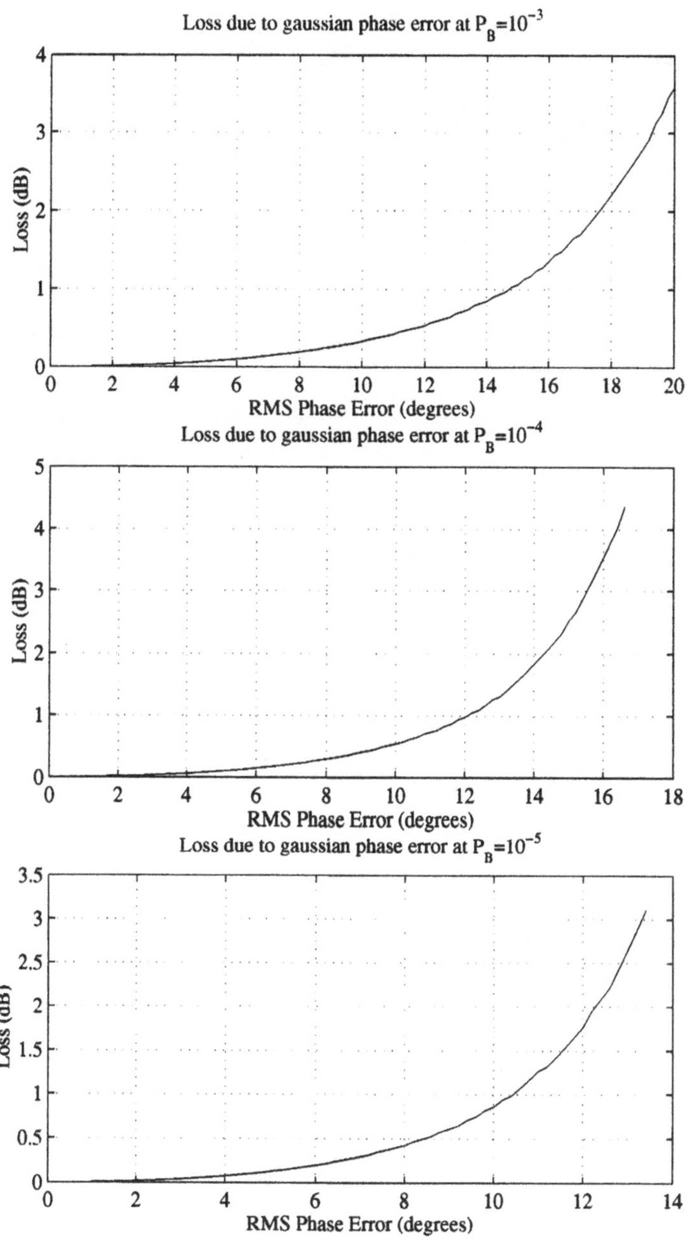

Figure 3.6. Signal to Noise Loss Due to a Gaussian Phase Error

this chapter is that the residual phase error should be less than approximately 10°_{RMS} to achieve reasonable bit error probability performance.

Chapter 4

AUTOMATIC CALIBRATION SYSTEM

In this chapter, we will discuss the block diagram for the automatic calibration circuit. The parameter which will be set by the automatic calibration is a gain constant in the form of a reference current used by the charge pump in the PLL. The overall gain in the PLL forward path depends on the absolute value of an integrated circuit capacitor and the slope of the VCO tuning curve. Neither of these parameters is tightly controlled. The basic overall block diagram in shown in Figure 4.1. The phase of the RF output, $\phi_{RF}(t)$, is sampled and quantized to produce $\phi_{RF}(kT)$ where T is the sample rate of the system. The RF output phase contains an $\omega_c t$ term corresponding to the center frequency of the RF carrier. This phase ramp can be computed digitally and subtracted from $\phi_{RF}(kT)$. This calculation is possible because the center frequency is locked via the PLL to the same reference clock used by the DSP section.

After the phase ramp is removed, the result is compared to the desired phase function, $\phi_{mod}(kT)$. The residual phase error, $\phi_e(kT)$ is used to determine how to adjust the charge pump gain in the PLL. A sample of $\omega_{mod}(kT)$ is required in addition to $\phi_e(kT)$ to correctly drive the gain controller. To see why this extra information is required, consider the case of a simple modulation index error. If the only information available is that ϕ_e is less than zero, we don't know if a "1" was transmitted and the modulation index was too small, or if a "0" was transmitted and the modulation index too large. However, knowledge of the commanded phase provides the necessary information to make this determination and correctly control the gain.

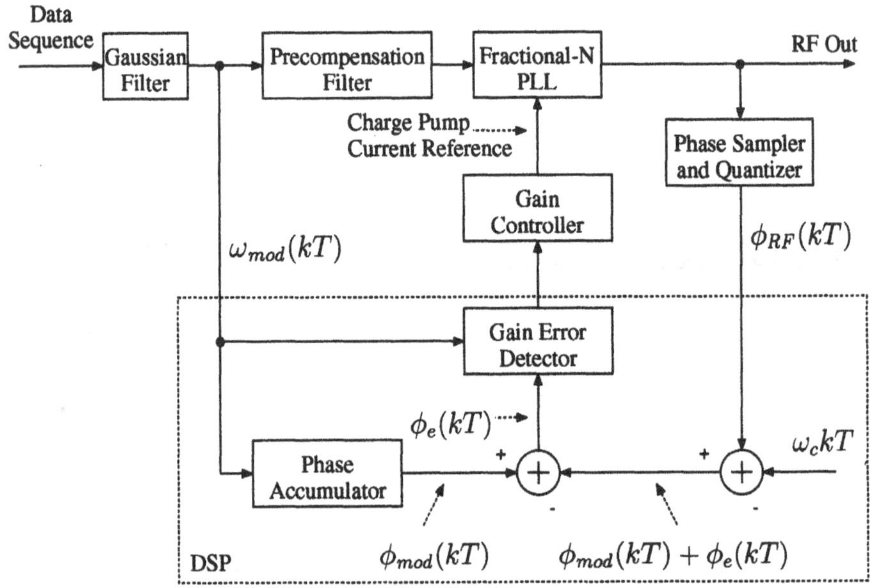

Figure 4.1. Automatic Calibration Circuit Overview

4.1. PLL Mismatch Model

To design the gain error detector, we must first examine how the PLL response varies with component value variation and how these variations show up in the output phase error. Specifically, we will see that the phase error at any time depends on a gain error and the history of what bits have been transmitted.

Figure 4.2 shows a simple model for the modulated fractional-N synthesizer. The divider has been linearized about its operating point. The nominal divide value is N and the variations about the operating point are $\tilde{n}$. All blocks are modeled as continuous time systems to simplify the analysis[1]. The phase detector gain, K_D is simply a scale factor relating phase error to duty cycle. Because of this K_D is fixed accurately. By using switched capacitor techniques in the loop filter, the time constants τ_z and τ_p may be tightly controlled. However, the gain term, $\frac{I_P}{C}$ which is the ratio of charge pump current to integration capacitor, is not very well controlled. The VCO tuning slope, K_{VCO}, also depends on absolute parameter values which are poorly controlled. Finally, the average divide

[1]Not all blocks are continuous time in nature, but for our purposes the continuous time model will suffice.

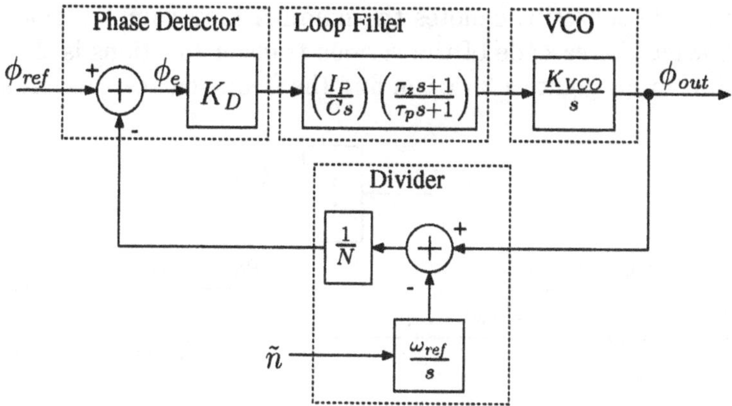

Figure 4.2. Modulated PLL Synthesizer

value, N, is digitally controlled but varies with center frequency. The conclusion is that the PLL response variation is controlled primarily by the gain constant. This conclusion has two consequences. There is only a single parameter we must adjust and since this parameter contains a current in it, we can easily adjust the gain.

The complete PLL open loop gain is

$$\begin{aligned} L(s) &= K_D\left(\frac{I_P}{Cs}\right)\left(\frac{\tau_z s+1}{\tau_p s+1}\right)\left(\frac{K_{VCO}}{s}\right)\left(\frac{1}{N}\right) \\ &= K_T\left(\frac{1}{s^2}\right)\left(\frac{\tau_z s+1}{\tau_p s+1}\right). \end{aligned} \tag{4.1}$$

The gain constant, K_T, may be written as

$$K_T = (1+\delta_k)K_{T_0} \tag{4.2}$$

where K_{T_0} is the nominal gain, δ_k is the fractional gain error and K_T is the actual gain.

The modulation transfer function of the PLL can be written as

$$H_{pll}(s) = \left(\frac{\Phi_{out}(s)}{\tilde{N}(s)}\right) = \left(\frac{-\omega_{ref}}{s}\right)\left(\frac{L(s)}{1+L(s)}\right) \tag{4.3}$$

The precompensation filter has a fixed transfer function based on the nominal characteristics of the PLL. The ideal precompensation transfer function is given by

$$H_{pre}(s) = \frac{1+L_0(s)}{L_0(s)} \tag{4.4}$$

where the "0" subscript denotes the nominal loop gain. A block diagram showing the cascade of the various transfer functions is shown in Figure 4.3. From Figure 4.3 we can find the frequency response from the

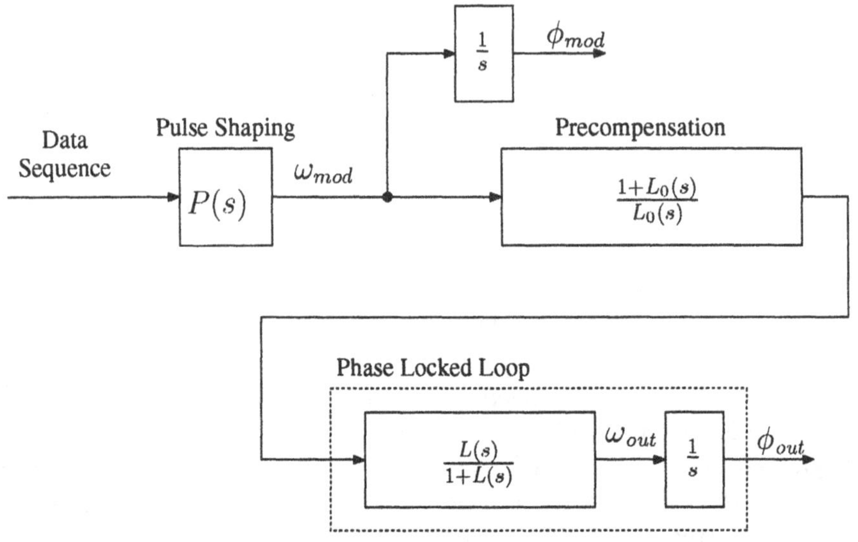

Figure 4.3. Mismatch Model

output of the pulse shaping filter to the modulation part of the instantaneous output frequency. The resulting transfer function, (4.5), has a DC gain of unity and a high frequency gain of $(1 + \delta_k)$.

$$H_{mod}(s) = \left(\frac{L(s)}{L_0(s)}\right)\left(\frac{1 + L_0(s)}{1 + L(s)}\right) \tag{4.5}$$

Making use of (4.1) we find

$$\left(\frac{L(s)}{1 + L(s)}\right) = \frac{\tau_z s + 1}{\frac{\tau_p}{K_T}s^3 + \frac{1}{K_T}s^2 + \tau_z s + 1}. \tag{4.6}$$

Under the assumption that τ_z and τ_p are tightly controlled, we can replace them with their nominal values, τ_{z_0} and τ_{p_0}, in (4.6). Performing this substitution and substituting (4.6) into (4.5) we obtain

$$H_{mod}(s) \approx \frac{\frac{\tau_{p0}}{K_{T_0}}s^3 + \frac{1}{K_{T_0}}s^2 + \tau_{z_0}s + 1}{\frac{\tau_{p0}}{K_T}s^3 + \frac{1}{K_T}s^2 + \tau_{z_0}s + 1}. \tag{4.7}$$

Figure 4.4 shows the frequency response defined by (4.7) for varying amounts of PLL gain error.

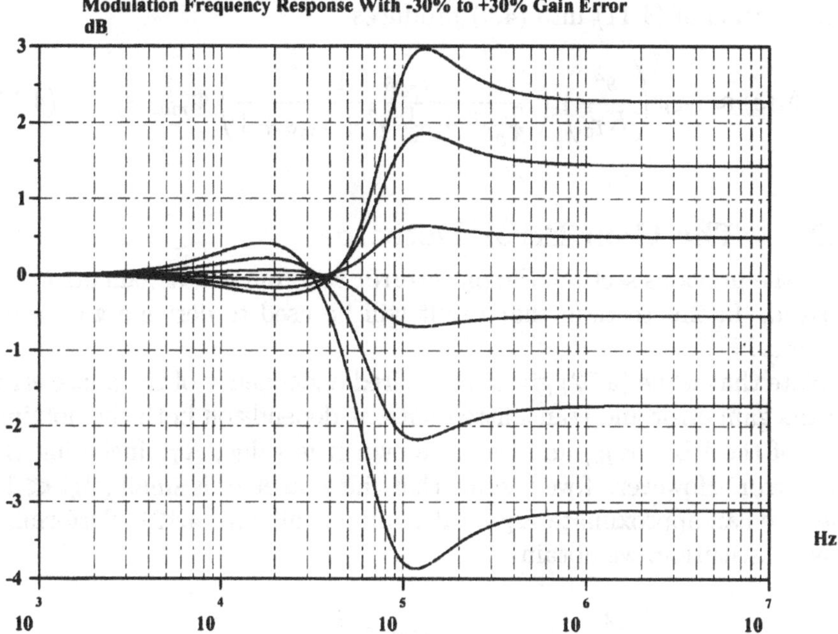

Figure 4.4. Precompensated Modulation Frequency Response with Mismatch

To understand the structure of the output phase error due to mismatch, we can calculate a transfer function between the desired output phase, Φ_{mod}, and the actual output phase, Φ_{out}.

$$\Phi_e(s) = \Phi_{mod}(s) - \Phi_{out}(s) = (1 - H_{mod}(s))\,\Phi_{mod}(s) \tag{4.8}$$

Substituting (4.5) into (4.8) yields

$$\Phi_e(s) = \left(\frac{\tau_{p0}\left(\frac{1}{K_T} - \frac{1}{K_{T_0}}\right)s^3 + \left(\frac{1}{K_T} - \frac{1}{K_{T_0}}\right)s^2}{\frac{\tau_{p0}}{K_T}s^3 + \frac{1}{K_T}s^2 + \tau_{z0}s + 1} \right)\Phi_{mod}(s). \tag{4.9}$$

The term $\left(\frac{1}{K_T} - \frac{1}{K_{T_0}}\right)$ may be written in terms of the gain error, δ_k, which gives

$$\left(\frac{1}{K_T} - \frac{1}{K_{T_0}}\right) = \left(\frac{1}{K_{T_0}}\right)\left(\frac{-\delta_k}{1+\delta_k}\right). \tag{4.10}$$

Assuming that $\delta_k \ll 1$ we can further simplify (4.10) to produce

$$\left(\frac{1}{K_T} - \frac{1}{K_{T_0}}\right) \approx -\delta_k\left(\frac{1}{K_{T_0}}\right). \tag{4.11}$$

Substitution of (4.11) into (4.9) produces

$$\Phi_e(s) \approx -\delta_k \left(\frac{s^2}{K_{T_0}}\right)\left(\frac{\tau_{p0}s+1}{\frac{\tau_{p0}}{K_T}s^3+\frac{1}{K_T}s^2+\tau_{z0}s+1}\right)\Phi_{mod}(s). \tag{4.12}$$

4.2. The Gain Error Detector

In the previous section, the phase error as a function of gain error was derived. In this section, that result will be used to derive a gain error detector.

Note that while (4.12) gives the desired dependency of the phase error on the gain error and transmitted data, it depends on both the nominal gain of the PLL, K_{T_0}, and the unknown true value, K_T, including the gain error. However, if we assume that the gain error is small ($|\delta_k| \ll 1$), then we can approximate K_T by it nominal value in (4.12). Performing this substitution, we obtain

$$\Phi'_e(s) = -\delta_k \left(\frac{s^2}{K_{T_0}}\right)\left(\frac{\tau_{p0}s+1}{\frac{\tau_{p0}}{K_{T_0}}s^3+\frac{1}{K_{T_0}}s^2+\tau_{z0}s+1}\right)\Phi_{mod}(s). \tag{4.13}$$

Noting that $\Phi_{mod}(s) = \frac{\Omega_{mod}(s)}{s}$, we can simplify (4.13) and obtain (4.14).

$$\Phi'_e(s) = -\delta_k \left(\frac{s}{K_{T_0}}\right)\left(\frac{\tau_{p0}s+1}{\frac{\tau_{p0}}{K_{T_0}}s^3+\frac{1}{K_{T_0}}s^2+\tau_{z0}s+1}\right)\Omega_{mod}(s) \tag{4.14}$$

The expression in (4.14) is very useful because it is made of a transfer function that we know *a priori* and a scalar multiplier which is the error term we are looking for. In particular, we can write ϕ_e as (4.15).

$$\phi_e \approx \phi'_e = \delta_k \phi_{e,ref} \tag{4.15}$$

The reference phase error signal, $\phi_{e,ref}$ is completely specified by the desired modulation frequency and a known transfer function and is given in (4.16).

$$\Phi_{e,ref}(s) = \left(\frac{s}{K_{T_0}}\right)\left(\frac{\tau_{p0}s+1}{\frac{\tau_{p0}}{K_{T_0}}s^3+\frac{1}{K_{T_0}}s^2+\tau_{z0}s+1}\right)\Omega_{mod}(s) \tag{4.16}$$

Figure 4.5 shows part of a possible sample path for $\phi_{mod}(t)$. Figure 4.6 shows the resulting $\phi_e(t)$, as calculated by (4.12), for various values of

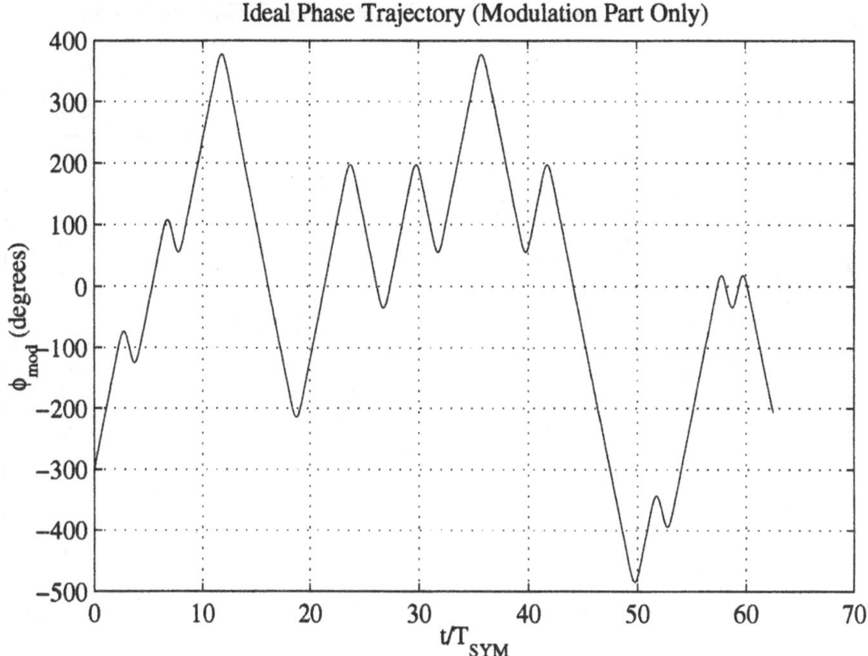

Figure 4.5. Sample Modulation Phase Trajectory

gain error. The figure also shows the reference error waveform, $\phi_{e,ref}(t)$. The figure shows that the general shape of $\phi_{e,ref}(t)$ closely matches the true phase error to within a scale factor as predicted by (4.15).

To determine the gain error, δ_k, we can do the following signal processing. Compute the reference error signal in (4.16). Measure the true phase error, $\phi_e(t)$. Now multiply the two signals. The result is (4.17).

$$\phi_e(t)\phi_{e,ref}(t) \approx \delta_k \phi_{e,ref}^2(t) \tag{4.17}$$

Examining (4.17) we see that the polarity of the detector output always matches the polarity of the gain error. In addition, the average of the detector output will be proportional to the amplitude of the gain error (assuming that the transmit bits are random). This output signal can be applied to an integrator whose output controls the forward path gain in the PLL.

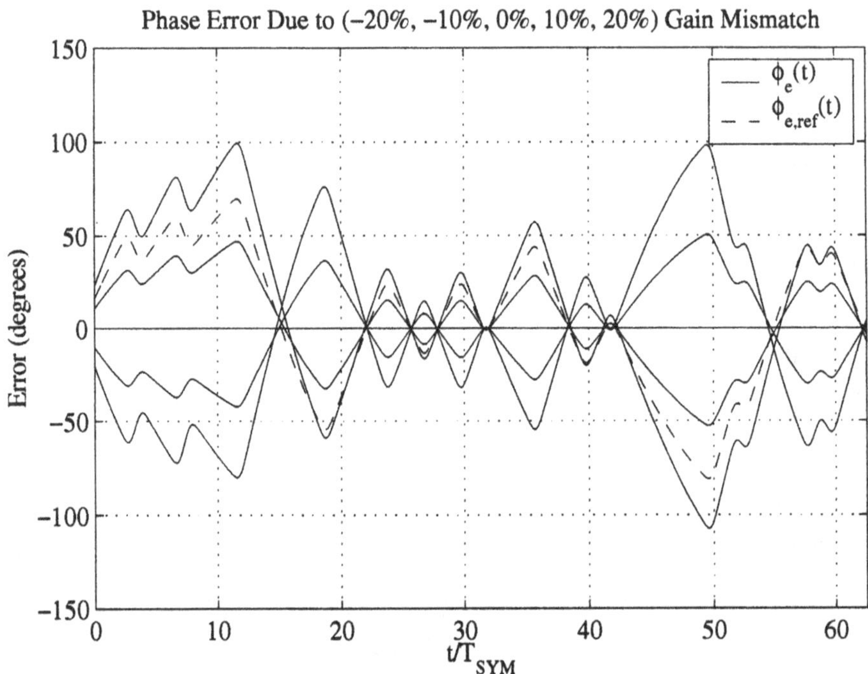

Figure 4.6. Sample Modulation Phase Error Trajectory

4.3. Track and Hold Modification to the Error Detector

While the gain error detector developed in section 4.2 looks promising, there are some important modifications which must be made for it to work. The circuit which measures the true output phase of the PLL will only provide an output in a 2π radian range. This modulo-2π reduction of the sampled, and also in the calculated, phase has a dramatic effect on the detector performance: it breaks it. To illustrate the effect, consider the case of two periodic signals with the same frequency, but with a fixed phase offset. We can write the phase functions of the two signals as $\phi_A = \omega t$ and $\phi_B = \omega t + \phi_{ofs}$. Clearly the difference between ϕ_A and ϕ_B is the constant ϕ_{ofs}. Now consider the modulo-2π reductions of the two phase functions. The reduced versions[2] are $\phi_{A,red} = ((\phi_A + \pi) \bmod 2\pi) - \pi$ and $\phi_{B,red} = ((\phi_B + \pi) \bmod 2\pi) - \pi$. Both $\phi_{A,red}$ and $\phi_{B,red}$ are zero

[2] By writing $\phi_{red} = ((\phi + \pi) \bmod 2\pi) - \pi$ we end up with a value in the interval $[-\pi, \pi)$ rather than $[0, 2\pi)$. The former is the required interval for the types of processing performed here.

mean functions and hence their difference is also zero mean. This means that the computed phase error is sometimes the same sign as the true phase error, sometimes the opposite sign, but on the average it is zero. This is illustrated in Figure 4.7.

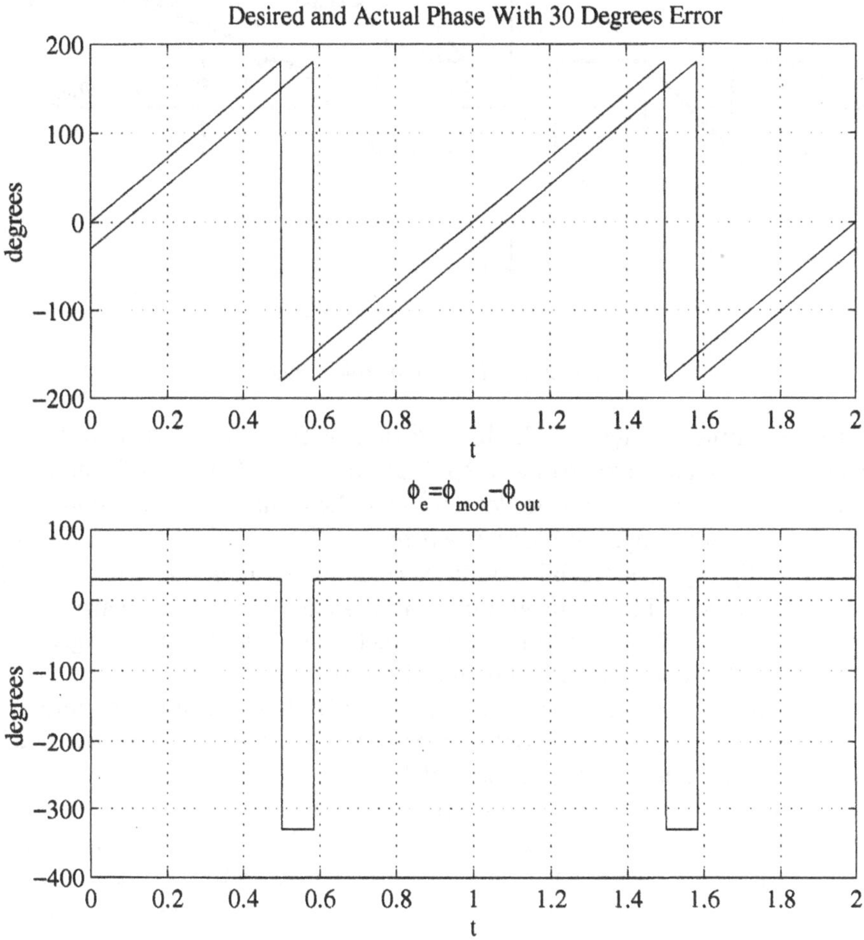

Figure 4.7. Phase Detector Waveforms–Fixed Phase Offset

Fortunately, a simple modification to the detector can restore its effectiveness. Consider a track and hold block which performs the following function. The input to the track and hold is $\phi_{out,red}(kT) - \phi_{mod,red}(kT)$ where $\phi_{out,red}(kT)$ and $\phi_{mod,red}(kT)$ are the sampled and modulo-2π reduced output phase and desired phase. When the input is between $-\theta$ and θ, the output is simply equal to the input. When the input is outside of that range, the output is made equal to the previous output. This action is illustrated for $\theta = \pi$ in Figure 4.8. For the constant phase

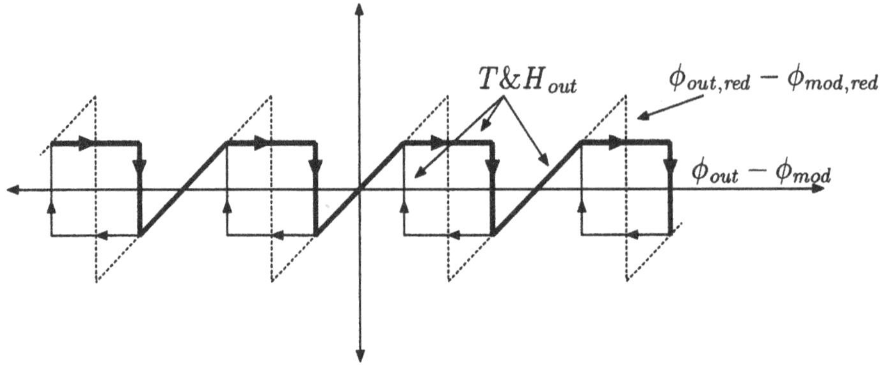

Figure 4.8. Track and Hold Detector

offset example in Figure 4.7, the addition of the track and hold causes the correct phase error to be output at all times. The use of track and hold blocks to convert phase detectors to phase and frequency detectors in phase locked loops was described by [26, 27].

As a slightly less trivial example, consider two signals with a fixed frequency offset. The reduced phases and associated difference are shown in Figure 4.9. The addition of the track and hold block changes the output in Figure 4.9 to that shown in Figure 4.10. We can see from these plots that the track and hold causes the output to contain a significant DC component of the same polarity as the true error. Without the track and hold, there is no DC component to drive an integrator.

The track and hold action described above can be added to the gain error detector to restore correct operation. The effect of a track and hold with $\theta = \pi$ on the gain error detector is illustrated in Figures 4.11, 4.12, 4.13, and 4.14. Figures 4.11 and 4.12 show the effect of the track and hold when the carrier frequency is set to an integer multiple of the reference frequency so that the carrier frequency portion of the sampled output phase is constant. Figures 4.13, and 4.14 are with the carrier frequency set to an integer plus a half multiple of the reference frequency so that the carrier frequency portion of the sampled output phase changes by π radians during each sample period.

4.4. Simulated Gain Error Detector Performance

Figures 4.15 and 4.16 show the gain error detector with track and hold modification. Figure 4.15 shows the system for measuring the output

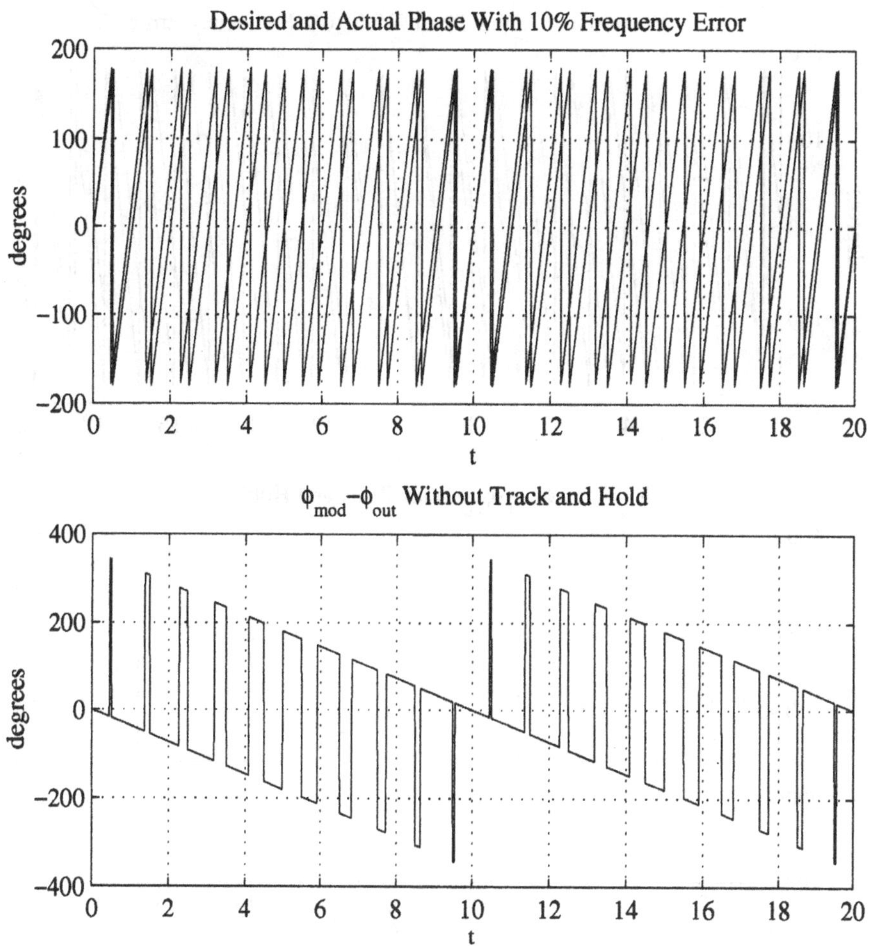

Figure 4.9. Phase Detector Waveforms–Fixed Frequency Offset

phase error. The desired phase modulation signal, ϕ_{mod}, is distorted by the PLL mismatch to produce ϕ_{out}. The $\omega_c kT$ term is due to the carrier frequency offset from an integer multiple of the reference frequency. The total RF output phase is then reduced to a 2π range and quantized. The desired phase modulation signal has $\omega_c kT$ added to it and the result is also reduced to a 2π range and quantized. This ideal sampled output phase is compared to the actual sampled output phase to produce the quantized phase error ϕ_{e_q}. In Figure 4.16, the quantized phase error, ϕ_{e_q}, is sent through the track and hold and the result is multiplied by

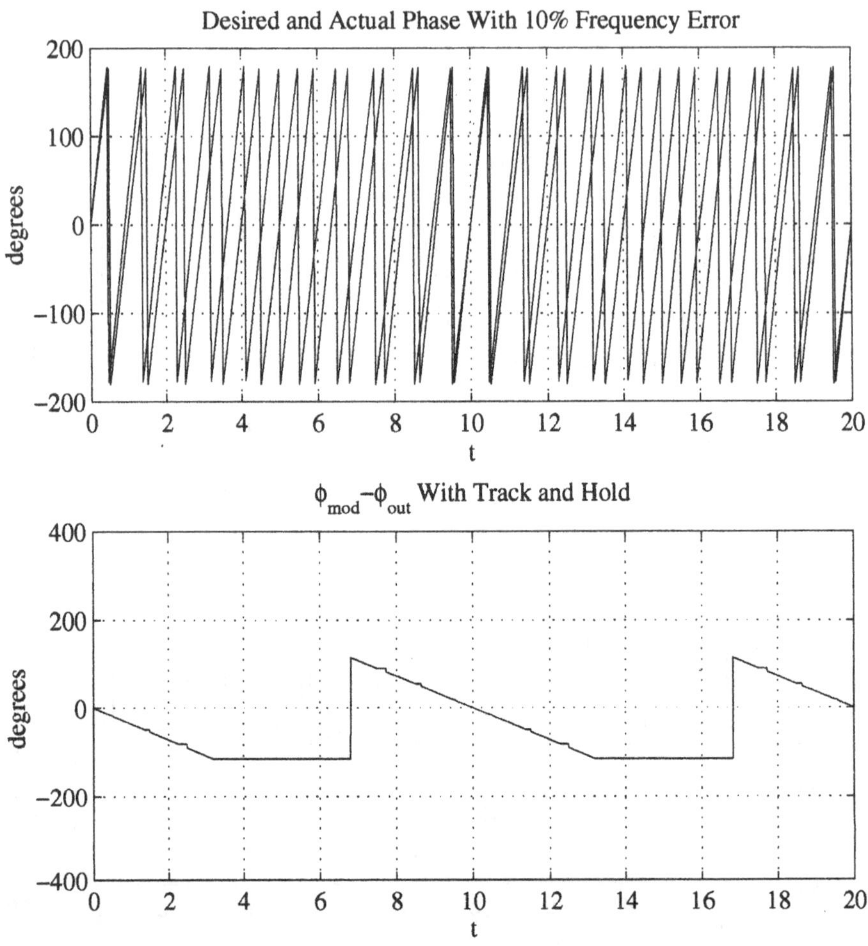

Figure 4.10. Track and Hold Detector Waveforms

the reference error signal, $\phi_{e,ref}$. The reference error signal is generated by filtering ω_{mod} in accordance with (4.16).

The detector has been simulated with 2 bit quantization of the output phase and also of the reference phase error signal. Figure 4.17 shows the average detector output versus gain error. The scale factor is chosen such that the maximum instantaneous output signal from the multiplication in Figure 4.16 is $\pm 100\%$. For gain errors in the range of approximately -50% to 75% the detector produces useful output. Figure 4.18 shows an expanded view of the detector output in the -5% to 5% gain error range.

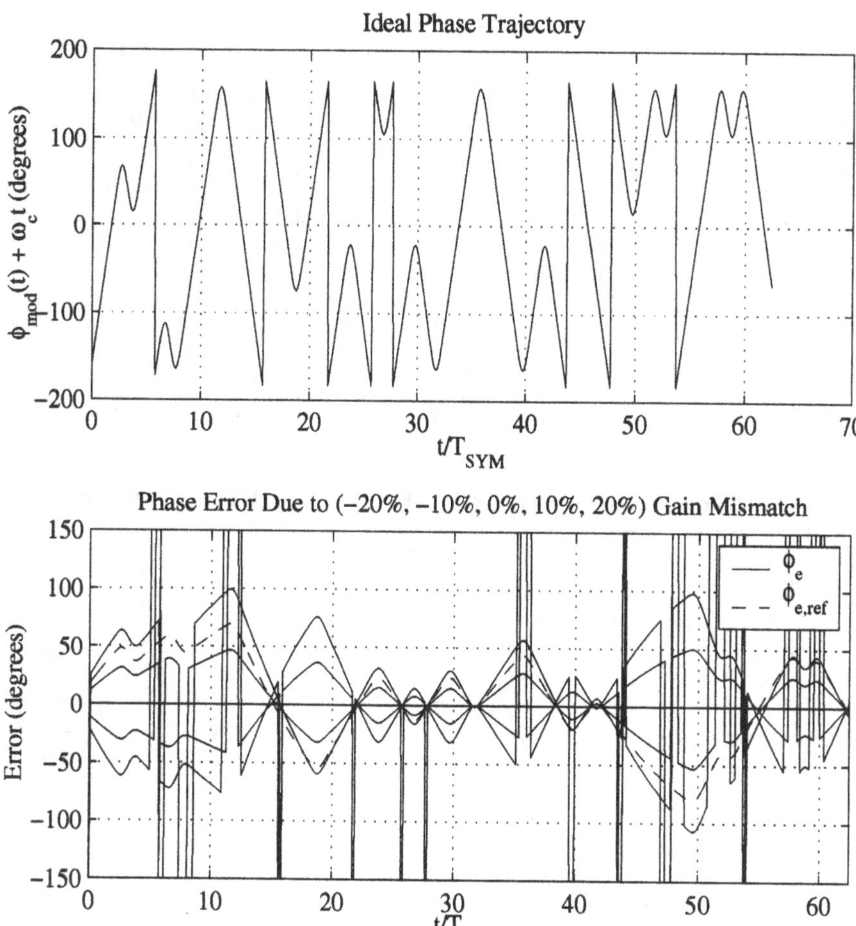

Figure 4.11. Phase Trajectory With Modulo–2π Reduction Without Track and Hold

4.5. Carrier Phase Tracking

4.5.1 Sources of Carrier Phase Error

In the previous sections it was assumed that the RF output phase as a function of time is precisely described by (4.18).

$$\phi_{RF}(t) = \phi_{out}(t) + \omega_c t \tag{4.18}$$

In (4.18), $\phi_{out}(t)$ is related to the transmitted data in some deterministic fashion and ω_c is the carrier center frequency. Since ω_c is set by the reference frequency and a digital divider, digital circuits which are clocked by the reference clock can have an exact knowledge of ω_c. These circuits

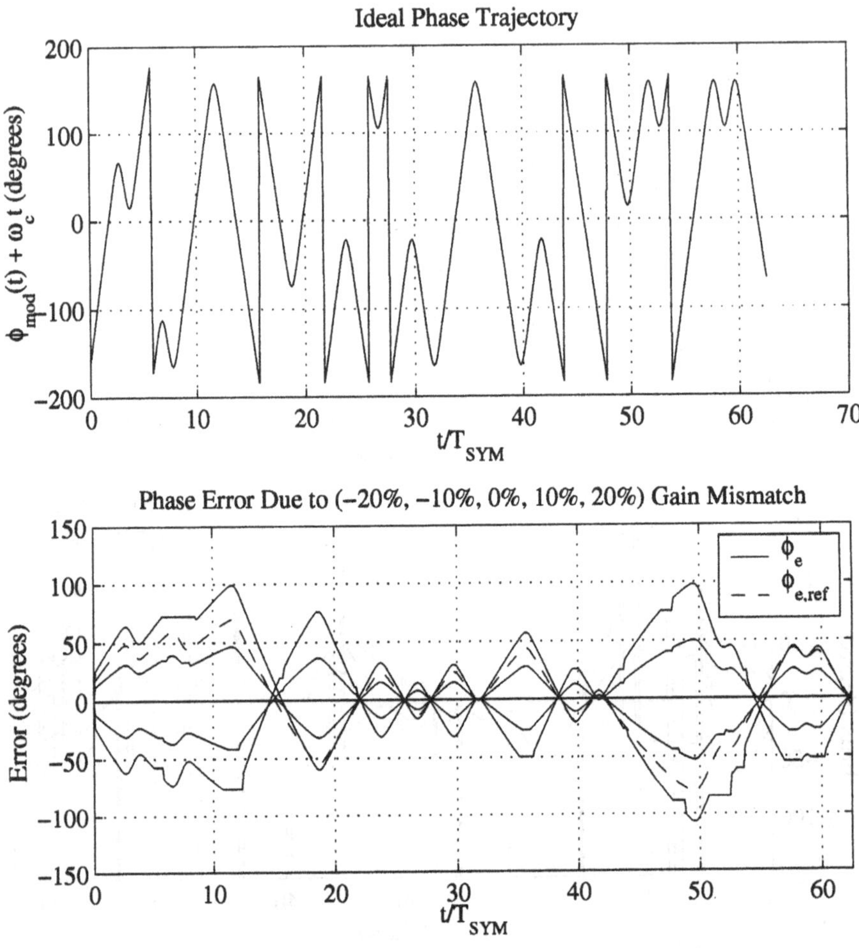

Figure 4.12. Phase Trajectory With Modulo–2π Reduction With Track and Hold

do not, however, have a zero phase reference. This simply means that there will be a fixed phase offset between the true carrier phase and a digitally calculated phase. An additional phase offset occurs due to the delay through the frequency divider and any delay associated with the RF front end of the phase quantizer.

In addition to the static phase offset, there will be a random drift between the sampled RF phase and the calculated phase. The source of this drift is the quantization in the transmit signal path and in the path which calculates the ideal output phase. The cause of this random drift is fairly simple. The PLL acts as a digital to frequency converter

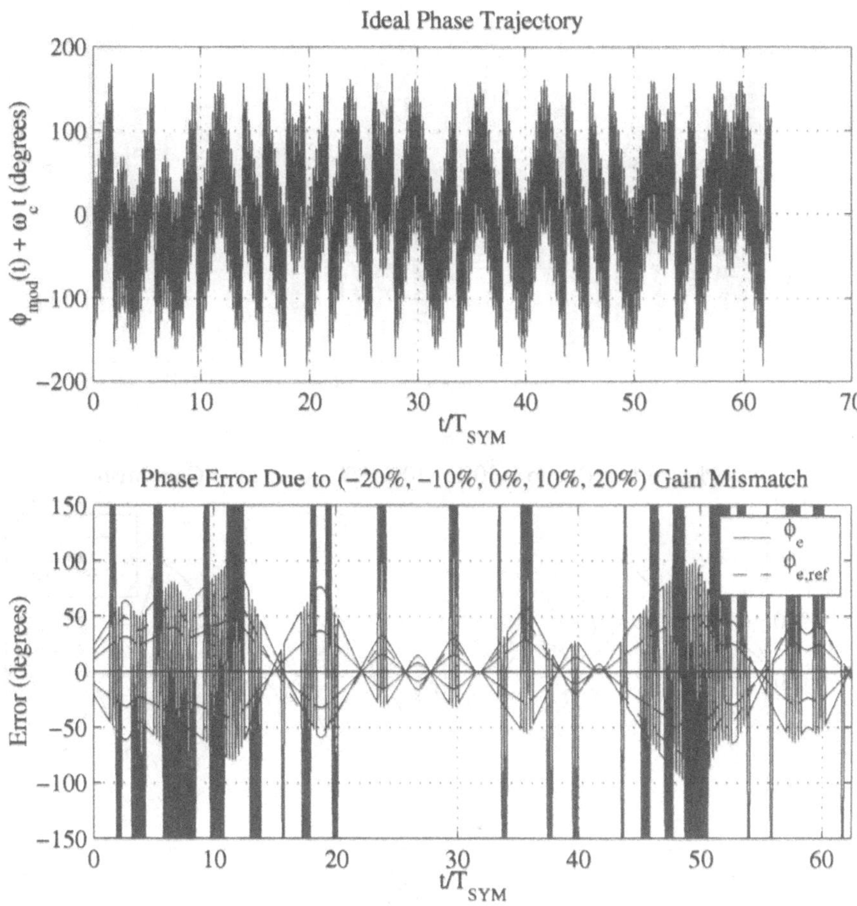

Figure 4.13. Phase Trajectory With Modulo-2π Reduction Without Track and Hold

and the output phase is the time integral of the output frequency. The ideal modulating signal is quantized during the pulse shaping and pre-emphasis filtering. The quantization error is then passed to the output frequency and integrated to produce a random walk in the output phase. Similarly, in the signal processing path that calculates the ideal output phase, there is a random walk due to the quantization. Since the quantization characteristics of the two signal paths are different, there will be a random walk in the difference between the measured and calculated phase.

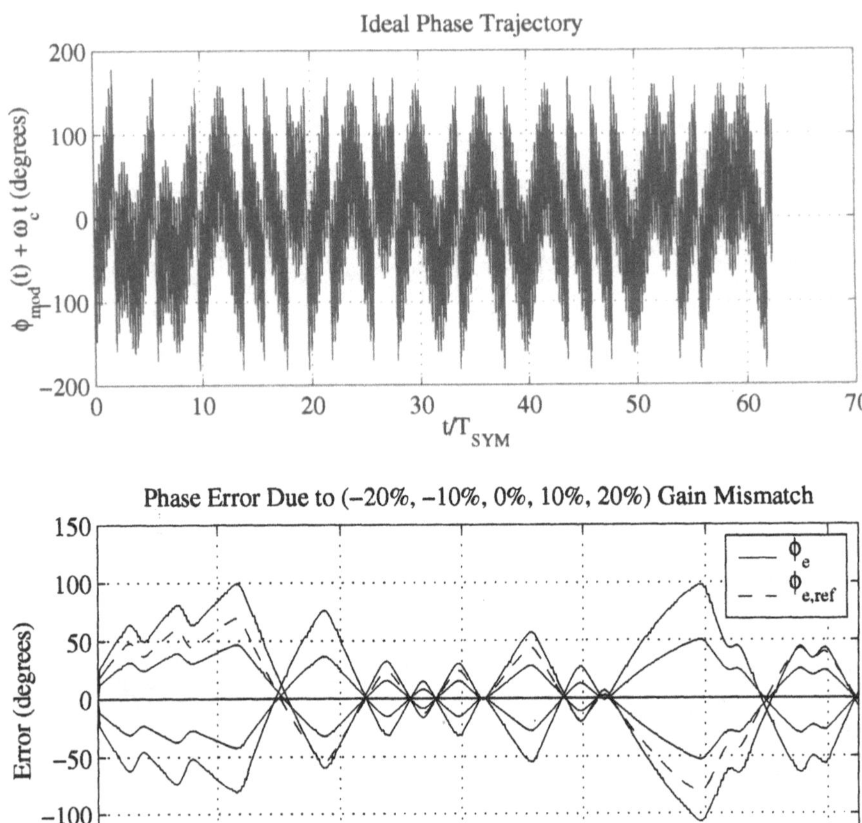

Figure 4.14. Phase Trajectory With Modulo–2π Reduction With Track and Hold

4.5.2 Effect of Phase Offset and Drift on the Gain Error Detector

At this point we need to decide if the presence of the static phase offset and the random drift is of concern. Figure 4.19 shows the effect of a static phase offset on the gain error detector output. We see that for relatively large phase offsets (up to 45 degrees) the detector works, but as the offset approaches 180 degrees the detector completely fails. Fortunately it is relatively simple to modify the phase error detector to remove the static phase offset and track the random drift.

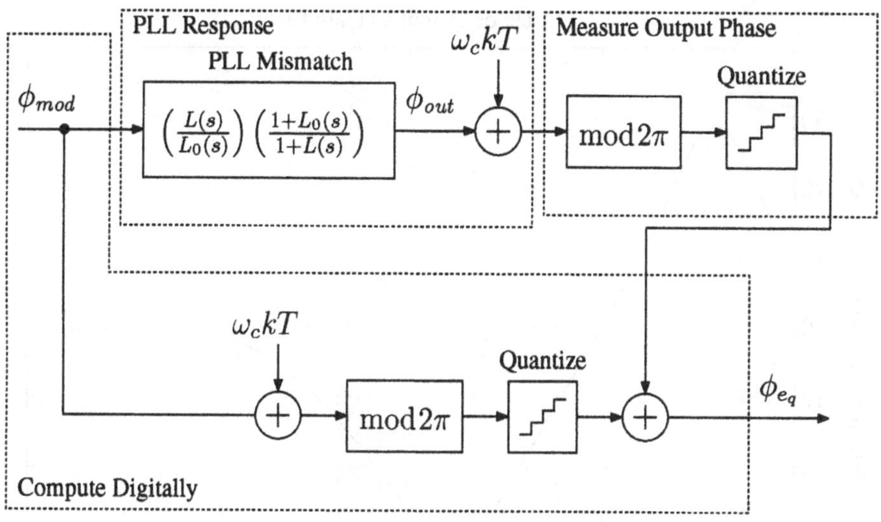

Figure 4.15. Output Phase Error Detector

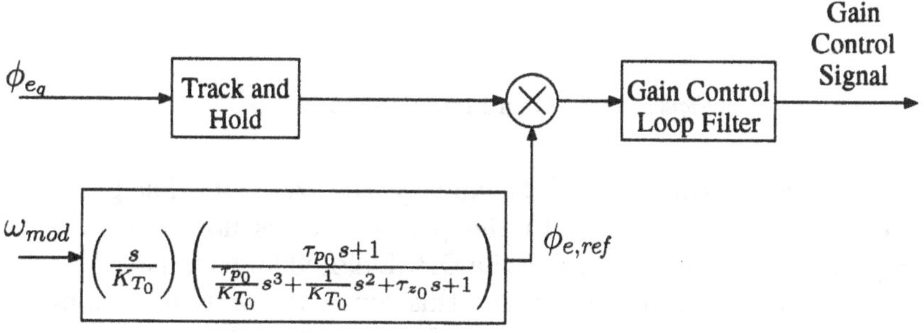

Figure 4.16. Gain Error Detector

4.5.3 Digital Phase Locked Loop Addition

The phase error detector part of the automatic calibration circuit is repeated in Figure 4.20. The phase accumulator in Figure 4.20 is allowed to wrap around when the value exceeds $\pm\pi$. In the absence of a phase offset, the detector output should have a time average value of zero assuming that the two possible transmit symbols have equal probability. We can use this fact to add an additional feedback loop which drives the average output to zero. Figure 4.21 shows the phase tracking addition.

The loop filter transfer function, $F(z)$, of the new digital phase locked loop (DPLL) should be designed to produce a DPLL loop bandwidth

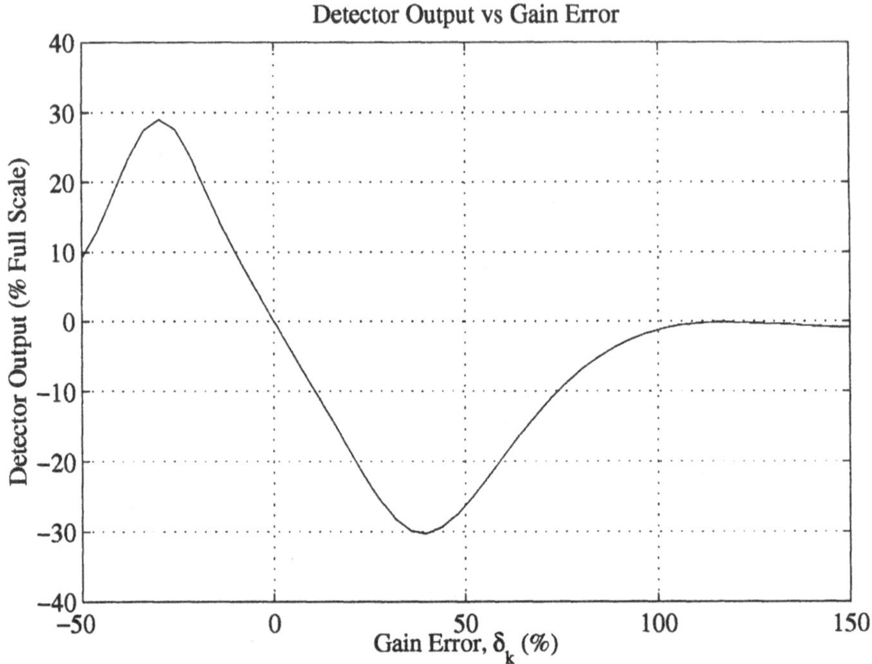

Figure 4.17. Gain Error Detector Output

which is much less than that of the main synthesizer loop. The previously derived response from ω_{mod} to the phase error is now multiplied by $1-H_{DPLL}(z)$, where $H_{DPLL}(z)$ is the DPLL closed loop transfer function. Above the DPLL loop bandwidth, this extra factor is approximately equal to one. Before the addition of the digital phase tracking, the spectral content of the phase error signal was more or less flat from 0 frequency to the main synthesizer loop bandwidth. By making the DPLL bandwidth relatively low, the total error signal level is largely unchanged. The DPLL is effective in removing the static phase offset and slow drift.

4.6. Complete Gain Error Detector

Figure 4.22 shows the complete gain error detector block diagram. The RF output feeds the RF phase quantizer to produce ϕ_{RF_q}. The quantized RF phase is compared to the ideal value which is generated by accumulating the ideal frequency modulating signal. The track/hold block serves the dual purpose role of providing phase/frequency detection for the DPLL and removing the deleterious effects of the 2π phase

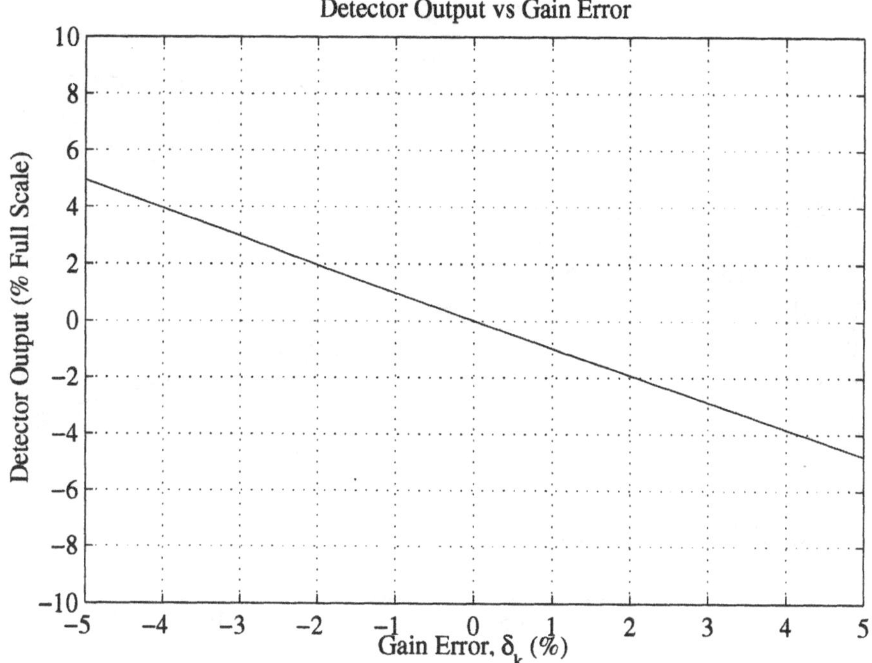

Figure 4.18. Gain Error Detector Output For Small Errors

ambiguity in the sampled and ideal phase. The slow digital phase tracking loop removes any static phase offset and random drift between the ideal and true RF phase. The reference error filter produces the reference phase error waveform which serves as the basis function for the measured phase error. By multiplying ϕ_e and $\phi_{e,ref}$ and lowpass filtering the result we obtain an estimate of the gain error. This estimate is used by the gain control error amplifier to set the gain and drive the error to zero.

4.7. Acquisition Range and Digital Coarse Calibration

The initial error that the calibration circuit can correct for is not limited by the curve in Figure 4.17 because the DPLL will fail to lock before the gain error detector output stops being useful. The amount of disturbance which is injected into the DPLL depends directly on the gain error in the main synthesizer. To estimate the capture range we first need to relate gain error to the total phase error injected into the

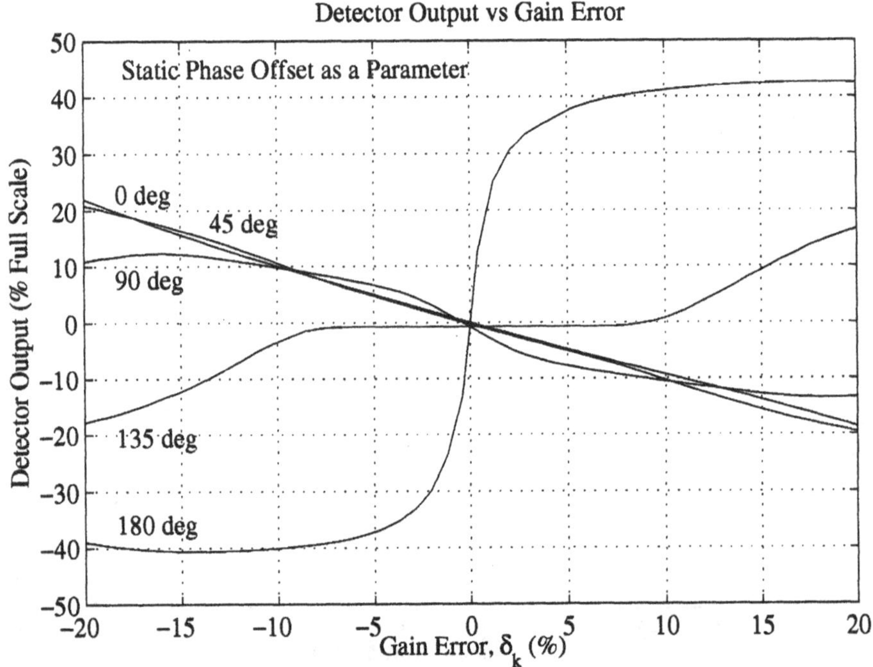

Figure 4.19. Gain Error Detector Output In the Presence of a Phase Offset

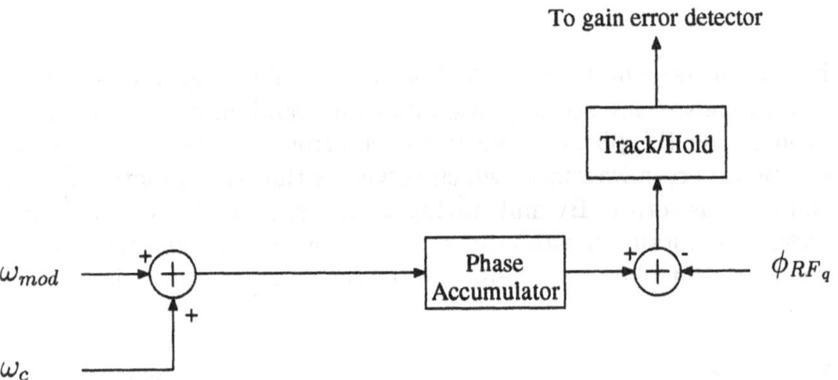

Figure 4.20. Phase Error Detector Without Tracking

DPLL. The phase error is given by (4.14), repeated here in (4.19).

$$\Phi_e'(s) = -\delta_k \left(\frac{s}{K_{T_0}}\right) \left(\frac{\tau_{p_0} s + 1}{\frac{\tau_{p_0}}{K_{T_0}} s^3 + \frac{1}{K_{T_0}} s^2 + \tau_{z_0} s + 1}\right) \Omega_{mod}(s) \qquad (4.19)$$

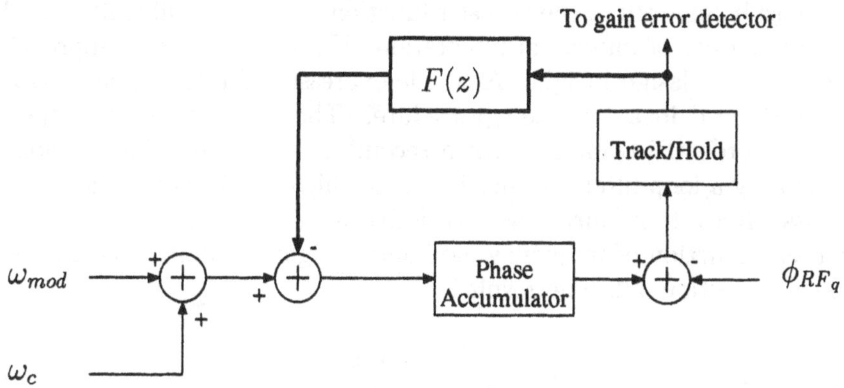

Figure 4.21. Phase Error Detector With Tracking

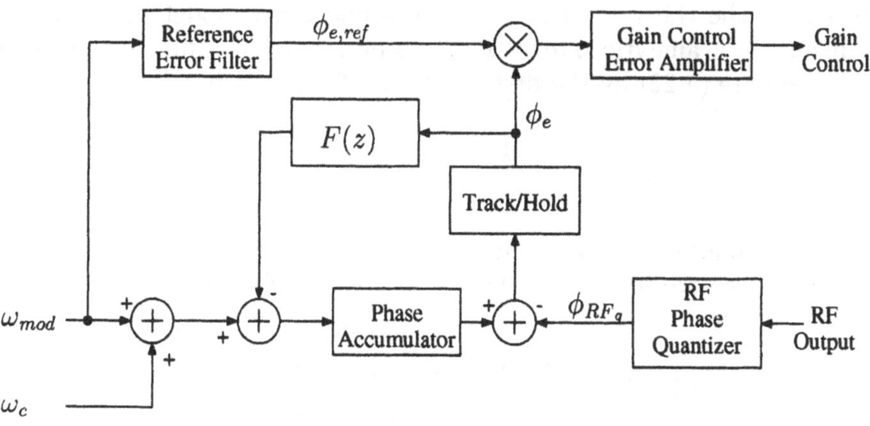

Figure 4.22. Complete Automatic Calibration Block Diagram

Since the bandwidth of the transfer function in (4.19) is much less than the symbol rate, we can approximate the frequency modulation by

$$\omega_{mod}(t) = \sum_k \frac{\pi}{2} a_k \delta(t - kT_{SYM}). \tag{4.20}$$

The corresponding (white) power spectral density is

$$S_{\omega\omega}(f) = \left(\frac{\pi}{2}\right)^2 \frac{1}{T_{SYM}}. \tag{4.21}$$

The noise bandwidth of the transfer function in (4.19) is difficult to find without the use of numerical integration. Unfortunately this approach provides little design insight. A simple expression for total phase error can be derived in a two step procedure. The first step is to replace the third order PLL model with a second order model. This is done by choosing a loop filter transfer function which consists of a first order lowpass filter. It is fairly easy to derive an expression for the phase error as a function of frequency modulation signal using the techniques applied in section 4.2. The result is

$$\Phi_e' \approx \left(\frac{-\delta_k}{Q_o\omega_{no}}\right)\left(\frac{Q_o\left(\frac{s}{\omega_{no}}\right)+1}{\left(\frac{s}{\omega_{no}}\right)^2+\frac{1}{Q_o}\left(\frac{s}{\omega_{no}}\right)+1}\right)\Omega_{mod}(s). \tag{4.22}$$

The natural frequency, ω_{no}, and loop quality factor, Q_o, are those corresponding to the complex pole pair in the third order model. While simpler than the third order model, the second order model still presents difficulties in analytic integration. To overcome this difficulty we can approximate (4.22) by the first order transfer function

$$\Phi_e'(s) \approx \left(\frac{-\delta_k}{Q_o\omega_{no}}\right)\left(\frac{1}{\frac{s}{Q_o\omega_{no}}+1}\right)\Omega_{mod}(s). \tag{4.23}$$

The total phase error variance is the product of the spectral density of ω_{mod}, the square of the DC gain, and the two sided noise bandwidth. The result is

$$\mathrm{var}(\phi_e) \approx \left(\frac{\pi}{2}\right)^2\left(\frac{1}{T_{SYM}}\right)\left(\frac{\delta_k}{Q_o\omega_{no}}\right)^2 2\left(\frac{\pi}{2}\right)\left(\frac{Q_o\omega_{no}}{2\pi}\right). \tag{4.24}$$

After some simplification, we obtain

$$\phi_{e_{RMS}} \approx \delta_k\frac{\pi}{2}\sqrt{\frac{f_{SYM}}{2Q_o\omega_{no}}}. \tag{4.25}$$

Figure 4.23 validates the approximations in the reduced order models by plotting the rms phase error as a function of gain error for the first, second, and third order models.

Note that the DPLL loop bandwidth is much less than ω_{no} and hence the total phase error within the DPLL is about the same as the input phase error. With a loop bandwidth of 81.4 kHz, a damping ratio of 0.77, and symbol rate of 2.5 Mbps, (4.25) becomes

$$\phi_{e_{RMS}} \approx \delta_k 175^\circ_{RMS}. \tag{4.26}$$

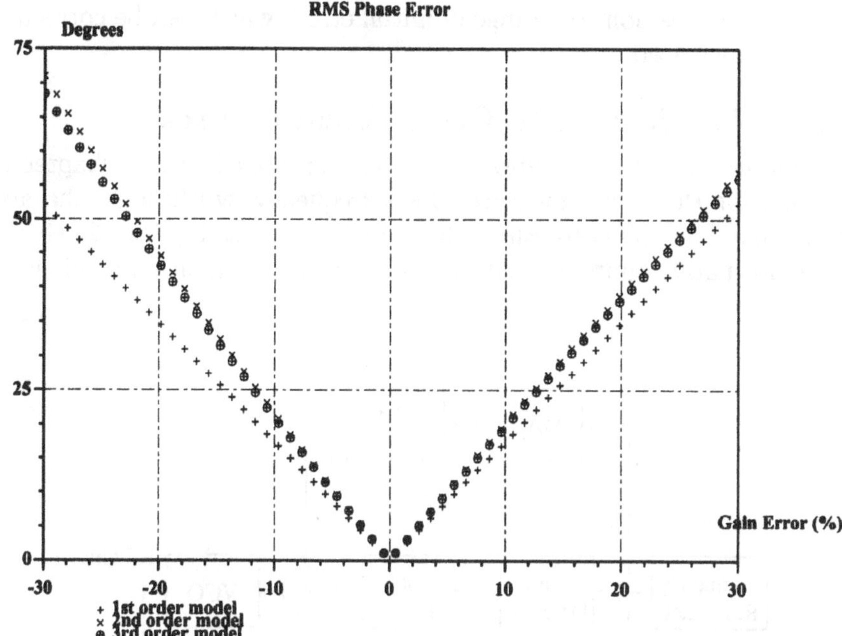

Figure 4.23. RMS Phase Error vs. Gain Error

The phase/frequency detector has a linear operating range of ±90°. Taking the peak phase error to be 6 times the RMS phase error, we obtain that the allowable gain error is about ±8.6%.

Unfortunately we expect the gain error due to process variation to be larger than the range found. A solution to this problem is to include a digital coarse calibration circuit which can cover a relatively large range of mismatch. The digital coarse calibration works in parallel with the automatic calibration circuit. The digital coarse calibration operates by monitoring the gain control error amplifier output. During acquisition, the coarse gain control DAC is stepped through its various settings. When the error amplifier output enters its normal output range, the digital sweep is turned off[3]. At this time, the digital logic should enter a track mode where the coarse calibration DAC is stepped by one value when it is detected that the error amplifier output leaves its nominal

[3]This sweeping type of acquisition aid is common in various phase locked loop applications. See [28] for example.

range. In this fashion, the range of initial errors which can be corrected is greatly increased.

4.8. Application to Other Architectures

While the automatic calibration system developed in this chapter is designed for the precompensated Σ–Δ frequency synthesizer, the approach may be applied to other related architectures. Figure 4.24 shows an an alternative approach [29] to extending the modulation bandwidth

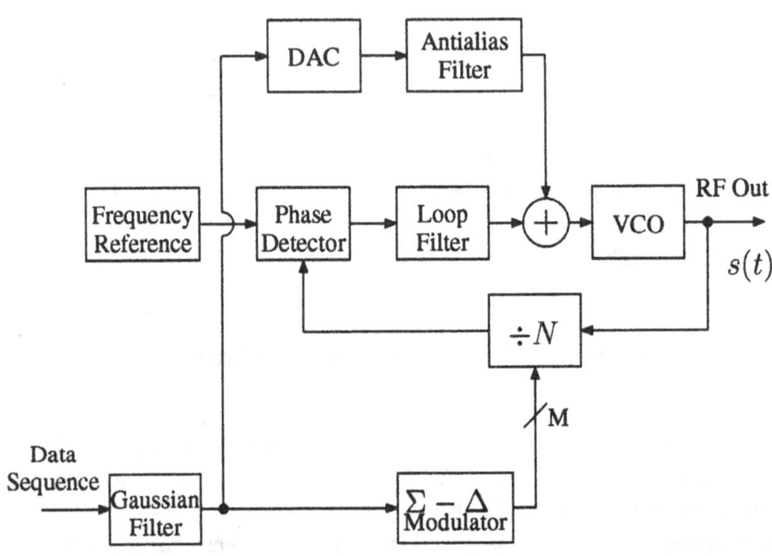

Figure 4.24. Split Path Modulated Synthesizer

of the Σ–Δ frequency synthesizer. The low frequency part of the modulation is delivered through the Σ–Δ modulator and divider while the high frequency portion of the modulation is delivered through the DAC. This architecture suffers from the same type of mismatch problems as the precompensated approach. The low frequency modulation path is precisely controlled via digital circuits while the high frequency path depends on absolute characteristics of the DAC and VCO. By including a variable gain element in the DAC, this architecture may be calibrated. To design the appropriate gain error detector, the approach described in this chapter may be used. The result will be a slightly different response for the reference error filter.

4.9. Chapter 4 Summary

This chapter has examined the relationship between the synthesizer forward path gain error, the modulation signal, and the RF output phase error. A simple block diagram has been developed for a system which is capable of estimating the gain error. The system does not interfere with normal operation of the modulator and hence can operate in the background. The output of the gain error detector will be used by the gain controller to tune the PLL response. A digitally controlled coarse calibration circuit can be used to extend the range of mismatch which can be corrected by the calibration circuit.

Chapter 5

IMPLEMENTATION DETAILS

A sigma–delta modulated synthesizer incorporating the new automatic calibration circuit has been implemented in two forms. The first implementation is a board level realization of the architecture. The output center frequency is around 60 MHz with a data rate of 78.125 kbps. The VCO is a discrete transistor design. All digital functions including the RF frequency divider were implemented in a single field programmable gate array (FPGA). The second implementation is in the form of a full custom BiCMOS integrated circuit. In the integrated circuit implementation, the RF output frequency is around 1.8 GHz and the data rate is 2.5 Mbps using 2 level GFSK and 5.0 Mbps using 4 level GFSK. This chapter presents the integrated circuit version of the transmitter.

5.1. Discrete Prototype

Figure 5.1 shows a photograph of the prototype system. The voltage controlled oscillator is a discrete transistor design. The RF phase quantizer is realized using an RC–CR quadrature network followed by a pair of ECL line receivers and ECL flip-flops. The charge pump is a discrete transistor design with two reference current inputs. One of the inputs comes from a coarse gain control DAC and the other from the automatic calibration circuit. The loop filter and gain control error amplifier use off the shelf op-amps.

All of the digital functions are implemented in a single Xilinx XC4010 field programmable gate array. The FPGA approach allows flexibility in the digital signal processing circuits. The carrier center frequency and all bandwidths are scaled down in frequency by a factor of 32 from the target frequencies used in the integrated circuit version. Using a low

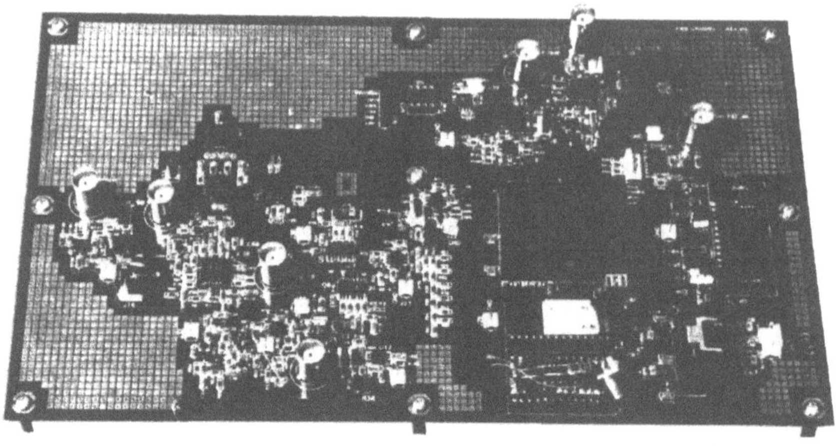

Figure 5.1. Photo of Low Frequency Prototype Board

carrier center frequency of around 60 MHz allows the implementation of the multimodulus divider in the FPGA as well.

A DAC is provided to monitor key digitized signals inside the digital signal processing circuitry. The remaining circuitry on the test board is a 10 MHz voltage controlled crystal oscillator (VCXO) that is used as the system reference. The VCXO may be phase locked to the 10 MHz reference output that is present on most Hewlett Packard RF test equipment.

5.1.1 Charge Pump and Loop Filter

The output of the phase/frequency detector in the main synthesizer is converted to a current which then drives the loop filter. Figure 5.2 shows the schematic of the discrete implementation of the charge pump and loop filter. The amplitude of the charge pump current is set in parallel by the automatic calibration circuit which controls $ICAL$ and the coarse digital calibration DAC which controls $IDAC$. Current mirrors are used so that the collector currents in Q6 and Q7 are both $ICAL+IDAC$. The sum of the collector currents in Q10 and Q11 is $2(ICAL+IDAC)$. The output current from the charge pump is then $\pm(ICAL+IDAC)$. The phase/frequency detector output, a TTL signal level signal, drives U1 which is a TTL→ECL level converter chip. The ECL output provides an ideal drive for the current switch formed by Q1 and Q2. The four transistors, Q4-7, in the PNP current mirror are in a single package which

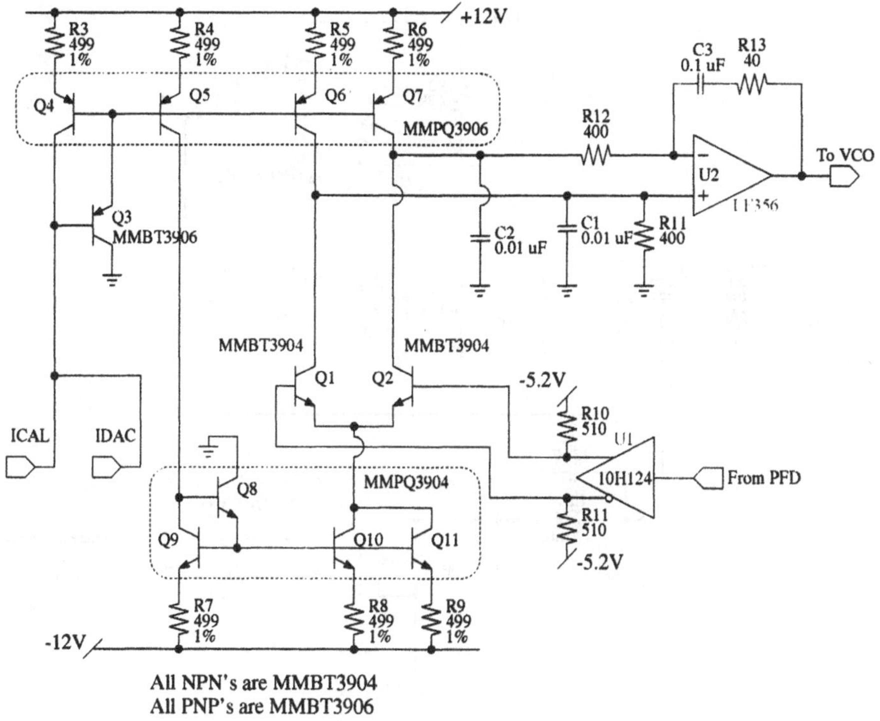

Figure 5.2. Discrete Charge Pump

improves matching and tracking over temperature. Similarly transistors Q8-11 in the NPN current mirror are in a single package.

The main synthesizer loop filter is formed by U2 and the associated resistors and capacitors. This particular topology was used to approximate the integrated circuit realization.

5.1.2 VCO

Figure 5.3 shows the schematic for the discrete prototype VCO. Transistor Q2 with feedback network C4 and C5 provides the negative resistance for the oscillator. The oscillator output is taken from the collector of Q2 to avoid disturbing the resonator. The common base amplifier, Q1, provides good reverse isolation. Since the current waveform in the oscillator resembles an impulse train rather than a sinusoid, a filter network comprised of C7-9 and L2 is used to produce a roughly sinusoidal output. The filter is a singly terminated elliptic type which was designed using the filter tables in [30]. An emitter-coupled-logic (ECL) line re-

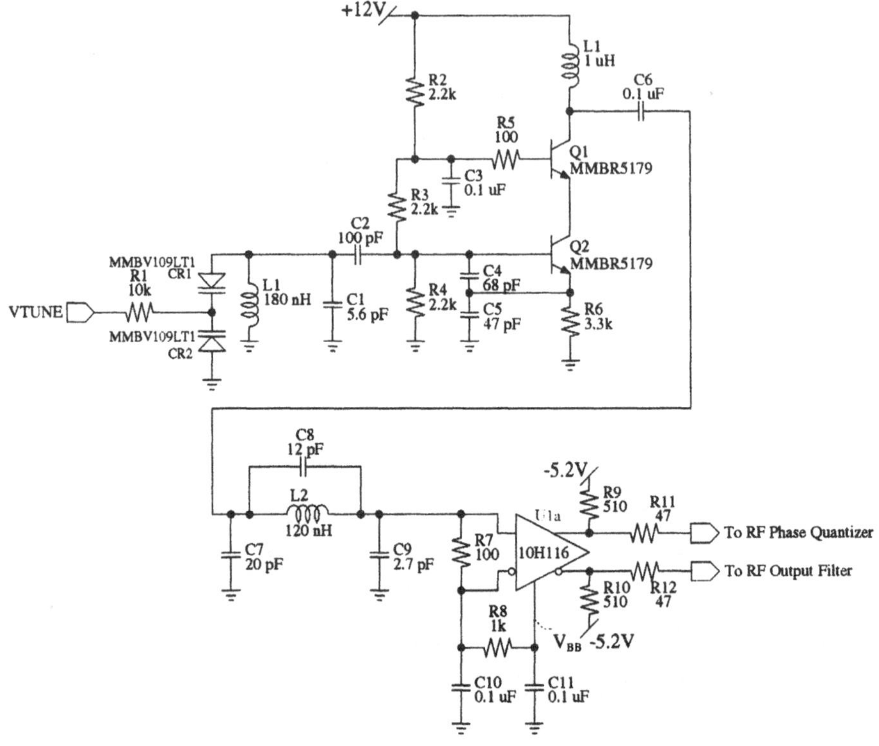

Figure 5.3. Discrete Voltage Controlled Oscillator

ceiver, U1a, is used to provide additional gain and present a 50 ohm impedance to the following stages. The bias voltage, V_{BB}, for the inputs of U1a is generated internally by the 10H116 and supplied to the inputs by R7 and R8.

5.1.3 RF Output Filter and Buffer

One of the two 50 Ohm outputs from the VCO buffer in Figure 5.3 is used to drive a RF bandpass filter and output amplifier. Figure 5.4 shows the RF bandpass filter and output amplifier. The two inductors, L1 and L2, are hand wound air core inductors and are used to tune the filter. The filter is a coupled resonator type [31], [32]. To tune the filter, a network analyzer is connected between the filter side of J1 and J3. A jumper is installed on J2 and L1 is adjusted until the resonator formed

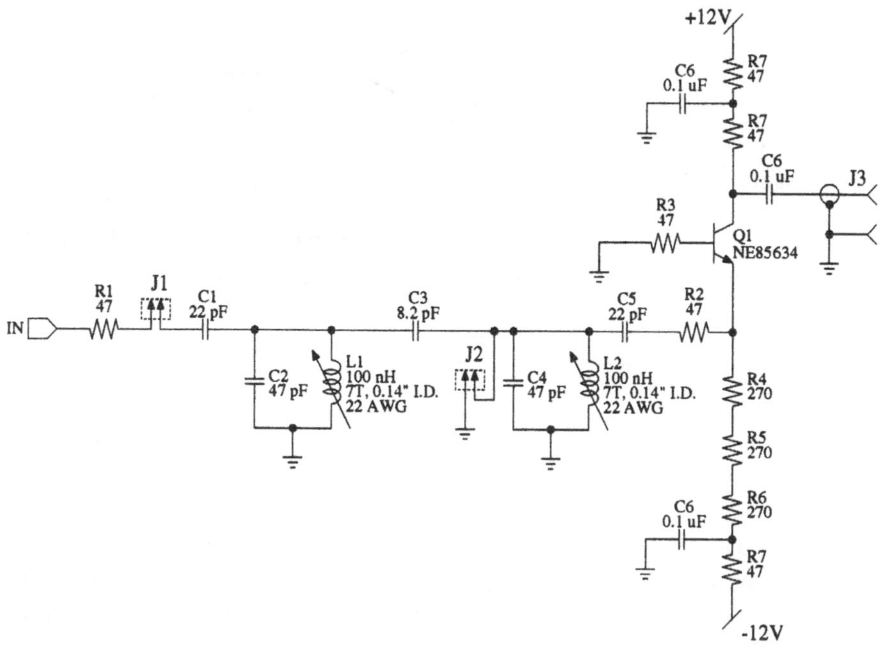

Figure 5.4. RF Output Filter and Buffer

by C2, C3, and L1 is resonant at the filter center frequency[1]. Then the jumper is removed on J2 and inductor L2 is adjusted until the desired transmission response is observed.

5.1.4 RF Phase Quantizer

An essential part of the automatic calibration circuit is the RF phase quantizer. The phase quantizer needs to operate with the RF carrier signal, around 60 MHz in the discrete implementation and 1.8 GHz in the integrated circuit implementation, as its input and only consume a small amount of power.

5.1.5 Quadrature Sampler Approach

The implemented phase quantizer in simplified form is shown in Figure 5.5. The modulated RF signal is passed through a quadrature network that provides two outputs whose phases differ by 90 degrees. The

[1] Resonance is measured with the network analyzer by plotting the input reflection coefficient on a Smith chart and adjusting L1 until the reflection coefficient angle is zero degrees.

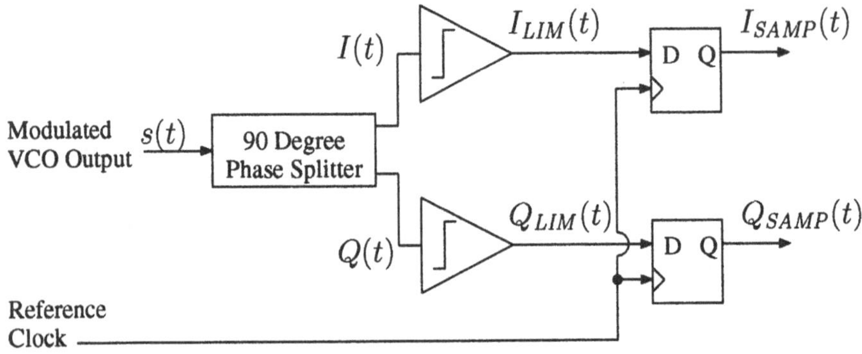

Figure 5.5. Phase Quantizer

quadrature signals are passed through limiting amplifiers and the results sampled by a pair of flip–flops. The flip–flop outputs now tell us which quadrant the RF phase was in at the sampling instant. The outputs

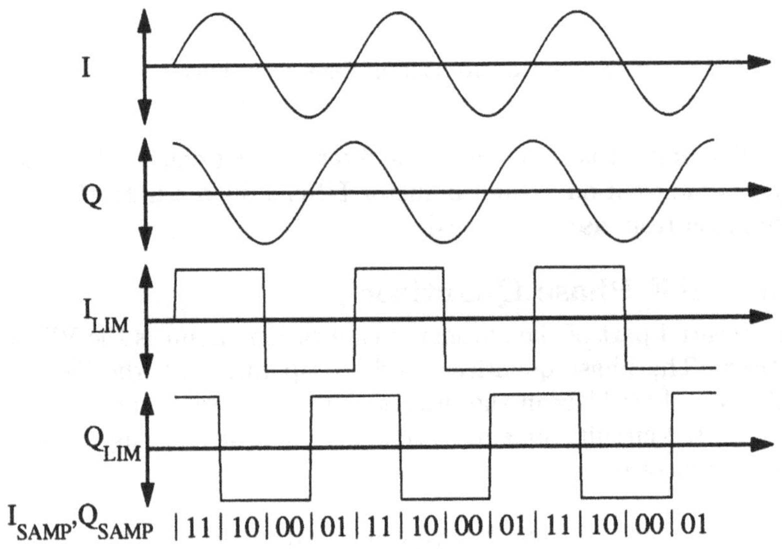

Figure 5.6. Idealized Phase Quantizer Waveforms

of the phase splitter and the limiters in the idealized case are shown in Figure 5.6. The binary values that would appear at the flip-flop outputs for various times of the reference clock rising edge are also shown. From the figure, we see that a 2π radian phase range is quantized to 2 bits.

Simulations show that this 2 bit quantization is sufficient for the gain error detector.

Figure 5.7 shows the schematic for a simple RC–CR 90 degree phase splitter. The outputs are always 90 degrees out of phase for sinusoidal

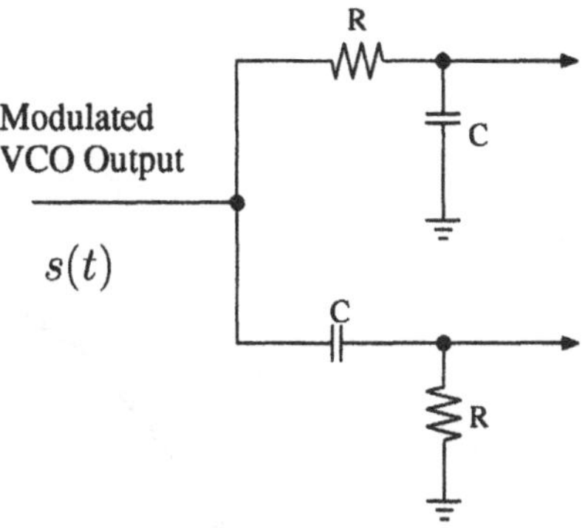

Figure 5.7. RC-CR Phase Splitter

inputs. In addition, the 2 output amplitudes are equal when $\omega = \frac{1}{RC}$. Mismatch between the time constants in the 2 arms of the phase splitter will cause a nonlinearity in the phase quantizer. The effect of this nonlinearity on the detector DC output, however, is minimal. Errors as large as 30 degrees are tolerable and that level of performance is easily achievable [33]. Figure 5.8 shows the effects of a quadrature error on the gain detector output. It is important to note that although we have introduced a quadrature phase splitter into the system, the performance requirements are much less than in the quadrature modulator. It is also important to note that the quadrature accuracy of the network in Figure 5.7 depends only on the matching between the two time constants and not their absolute values.

This type of phase quantization avoids the aliasing problem which would occur if the RF output were simply subsampled in a more traditional fashion. If a standard analog-to-digital converter were used to sample the RF output, then when the fractional frequency is near half of the reference frequency there would be aliasing in the digital output. The implemented phase quantizer essentially performs phase demodulation as part of the sampler[2]. The result is that the modulation portion of the sampled phase experiences no aliasing. The fractional frequency

[2] The sampler can be viewed as an I-Q downconverter followed by 1 bit quantizers and a digital phase demodulator.

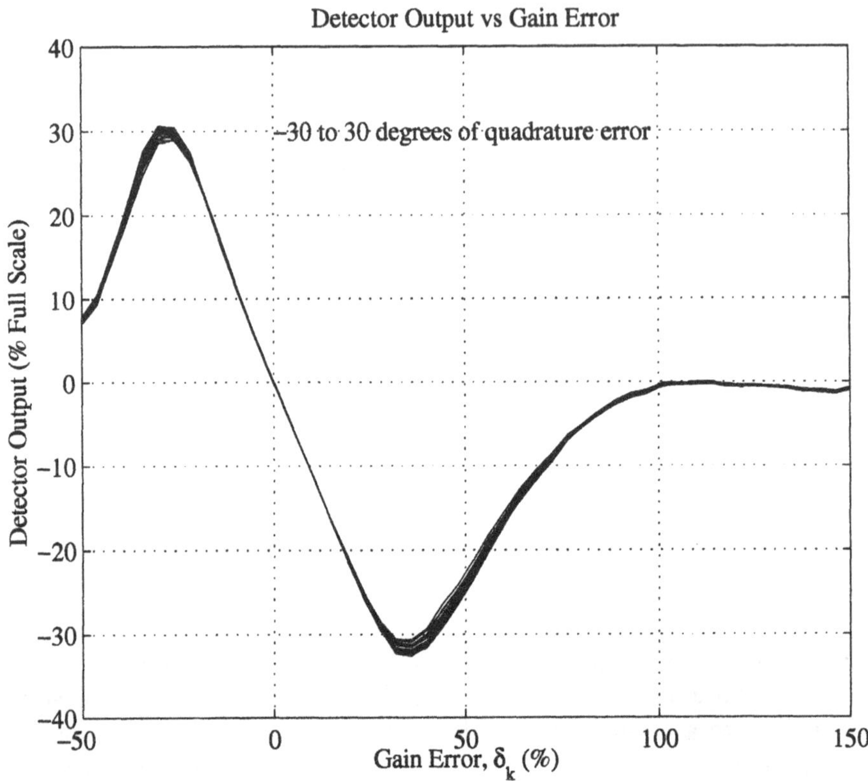

Figure 5.8. Gain Error Detector Output With 0° to 30° Quadrature Error

portion of the sampled phase does in fact experience aliasing, but this portion of the sampled phase signal is known and subtracted off.

5.1.6 Board Level Phase Quantizer Circuit Details

The second 50 Ohm output from the VCO buffer in Figure 5.3 is used to drive the RF phase quantizer. The RF phase quantizer is based on the quadrature sampler discussed above.

Figure 5.9 shows the schematic for the phase quantizer. The doubly terminated elliptic filter formed by C2, C3, C4, and L1 suppresses the harmonics which are present at the output of the VCO buffer. The quadrature network consists of the C-R section formed by C5 and R4 and the R-C section formed by R5 and C6. The elliptic filter needs to be terminated in a constant resistance for correct operation. The addition

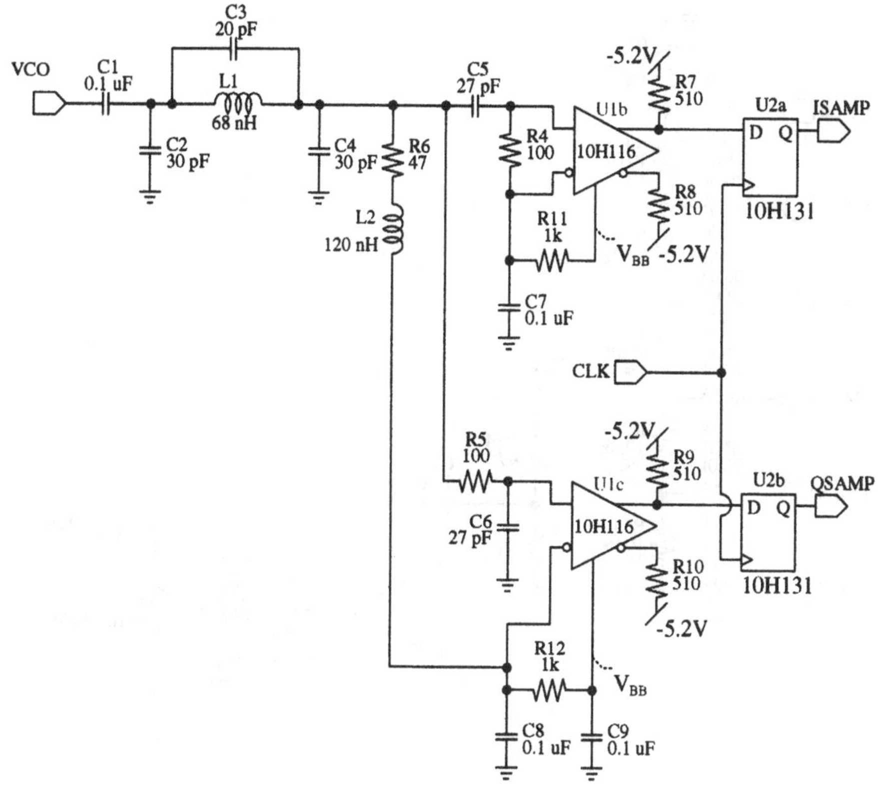

Figure 5.9. Discrete RF Phase Quantizer

of R6 and L2 to the quadrature network presents a constant 50 Ohm resistive termination for the filter.

The two outputs from the quadrature network drive ECL line receivers, U1b, and U1c, which are used as limiting amplifiers. The required input bias voltage, V_{BB}, is generated internally by the 10H116 and connected externally. A dual flip-flop, U2, samples the outputs from the limiters.

5.1.7 Digital Calibration DAC

To extend the range of the calibration circuit, a DAC is used to provide a coarse digital calibration. Figure 5.10 shows the schematic of the coarse calibration DAC. The DAC is a discrete R-2R ladder circuit. The $CAL_{5:0}$ inputs come from the CMOS outputs of the FPGA. The one percent resistor tolerance used in the ladder is sufficient to guarantee that the DAC is monotonic. The small offset injected by R13 at the

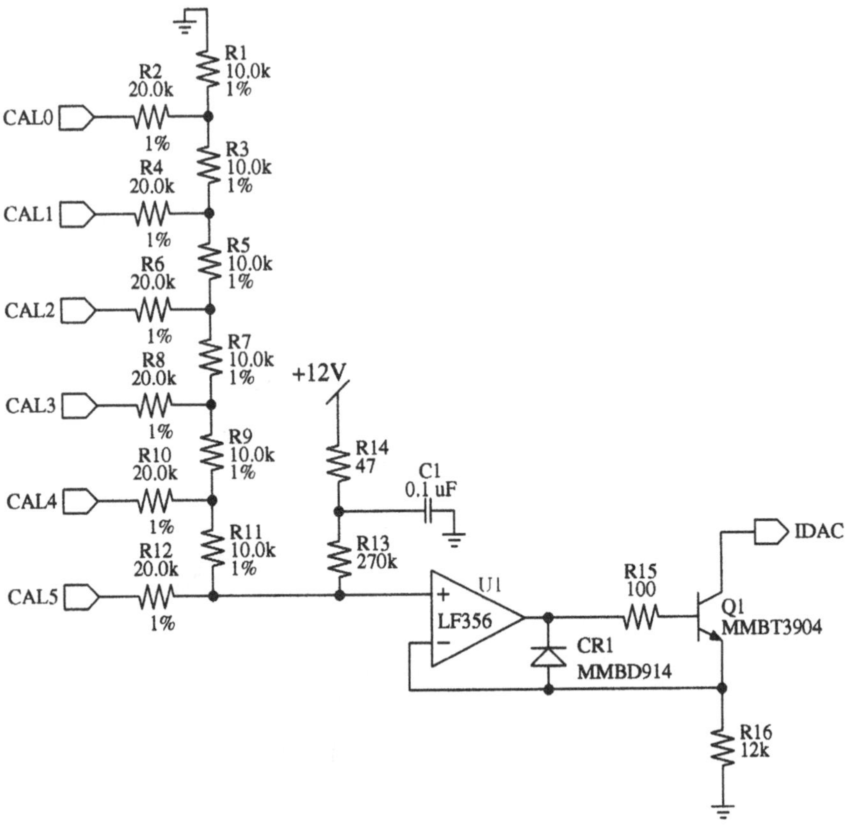

Figure 5.10. Discrete Calibration DAC

output of the ladder is to ensure that the input offset voltage of U1 doesn't cause the opamp output to swing negative. The R-2R ladder output voltage is converted to a current by U1, Q1, and R16.

5.1.8 Multimodulus Divider

The multimodulus divider used by the main synthesizer is required to provide a divide range of 64-127 in integer steps. This divider was realized as part of the Xilinx field programmable gate array. Figure 5.11 shows the FPGA schematic for the multimodulus divider. The design is based on the integrated circuit presented in [34]. The divide value is set by the signal $DIV_{5:0}$. The complete divider operates by allowing each of the individual divider stages to selectively swallow one input clock

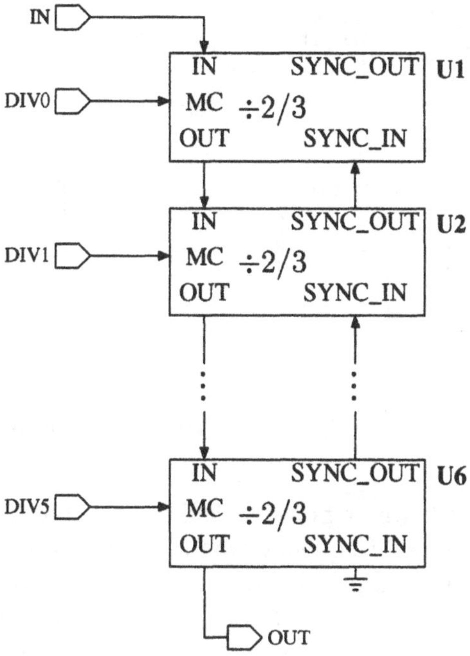

Figure 5.11. FPGA Implementation of a Multimodulus Divider

cycle[3] per output cycle of the complete divider chain. The selection is made by the individual modulus control inputs. The sync signal chain is used to notify each divider circuit when an output cycle has completed. This pulse swallowing action allows 0 to 63 input cycles to be swallowed per output cycle which produces the 64 to 127 divide range.

Figure 5.12 shows the FPGA schematic for the individual $\div 2/3$ stages. When the modulus control signal, MC, is low, the output of U2 is always low. In this state, U3 acts as an inverter and the combination of U3 and U4 forms a divide by two stage. In addition, the divide circuit always divides by two when $SYNC\_IN$ is high. When both MC and $SYNC\_IN$ are low, flip-flop U2 extends the output pulse by 1 clock cycle which produces 1 output cycle for every three input clock cycles.

5.1.9 Reference Oscillator

All of the clocks on the test board are derived from a 10 MHz voltage controlled crystal oscillator (VCXO). The VCXO can either be driven

[3] In this case "input clock cycle" refers to the input to that particular stage in the divider.

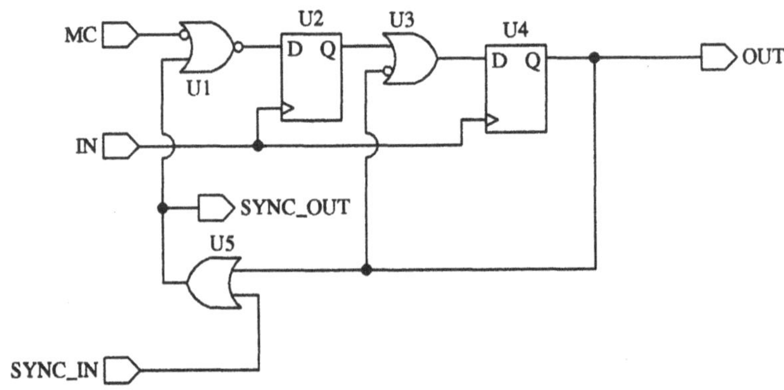

Figure 5.12. FPGA Implementation of a Divide by 2/3

by a fixed tuning voltage or phase locked to an external 10 MHz signal. This choice of reference frequency was made because several pieces of Hewlett-Packard test equipment use a 10 MHz reference. Typically HP test equipment will provide a sample of its frequency reference at a rear panel connector and also accepts an external reference. Having the board phase locked to the test equipment sometimes simplifies some of the measurements.

Figure 5.13 shows the schematic for the VCXO. An ECL line receiver is used as the gain element. The 10H116 used is essentially just a differential amplifier with emitter follower outputs. The crystal is cut to have its series resonant frequency be at 10 MHz. The series resonant circuit formed by the varactors, CR1 and CR2, and L1 is able to slightly shift the resonant frequency of the complete feedback network and tune the oscillator.

5.1.10 Digital Signal Processing

The digital signal processing which is part of the automatic calibration circuit was all implemented in the FPGA. The only difference between the FPGA implementation and the custom integrated circuit implementation is that the FPGA design was not pipelined. For further details on the digital phase locked loop and reference error filter, refer to the integrated circuit description.

5.2. Integrated Circuit Process Technology

The integrated circuit was designed and fabricated in a 0.6 micron BiCMOS process. Figure 5.14 shows the die photograph for the integrated circuit. The process includes two polysilicon layers allowing the

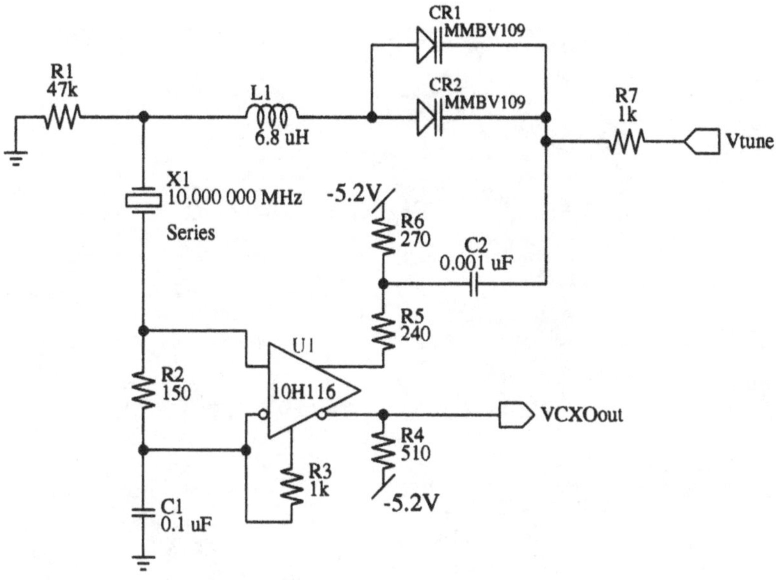

Figure 5.13. Voltage Controlled Crystal Oscillator

Parameter	*Symbol*	*Nominal Value*	*Units*
NMOS Threshold Voltage	V_{Tn}	0.72	Volts
PMOS Threshold Voltage	V_{Tp}	-0.91	Volts
Gate Oxide Thickness	t_{ox}	150	Å
Minimum Drawn Gate Length	L_{min}	0.6	μm
Poly1–Poly2 Capacitance	C_{PP}	0.45	$\frac{fF}{\mu m^2}$
Base Poly Resistor Resistance	R_{BP}	120	$\Omega/\Box$
Vertical NPN Peak Unity Current Current Gain Frequency	f_T	21	GHz

Table 5.1. Key Technology Parameters

use of poly–poly capacitors and also includes two metal layers. The vertical NPN devices feature peak values of f_T of more than 20 GHz. Key parameters of some of the supported devices are listed in Table 5.1.

In addition to the devices normally supported by the process, a varactor and bondwire inductor were included in the design. These additional devices do not represent any additional processing steps.

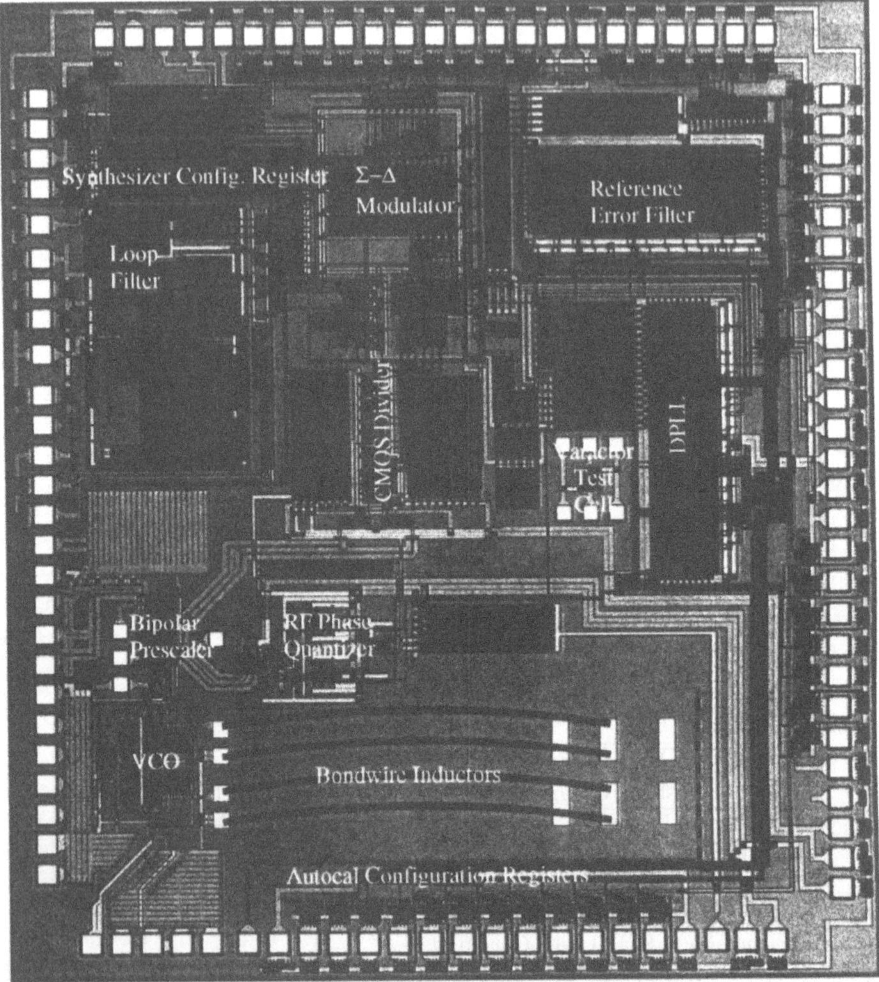

Figure 5.14. Modulator IC Die Photograph

5.2.1 Varactor

The varactor is formed by the base–collector junction of the supported vertical NPN transistor. To reduce parasitic series resistance, the emitter of the device is omitted and replaced by contacts to the base polysilicon. This layout results in a device like the one used in [35].

Figure 5.15 shows the layout of one of the varactor cells. Each varactor used in the design is made up of an array of 32 unit cells in parallel as

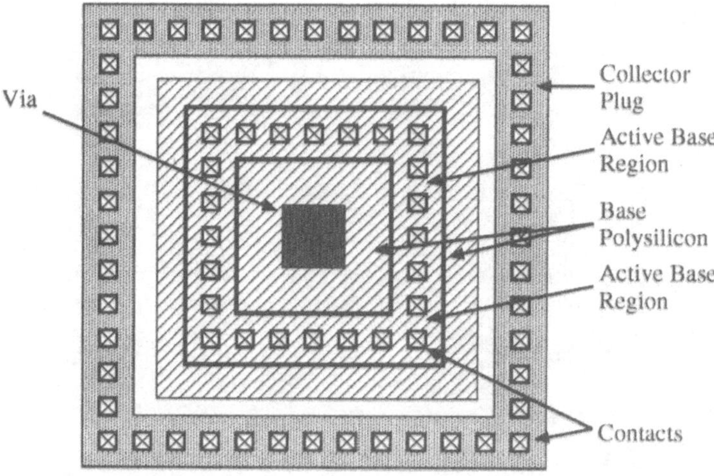

Figure 5.15. Varactor Unit Cell Layout

shown in Figure 5.16. The varactor was characterized using microwave wafer probes and extracting capacitance from the calibrated two port s–parameter data. The test chip layout includes a varactor test cell for these measurements. Figure 5.17 shows a photograph of the varactor test cell. Figure 5.18 shows the model used for the capacitance extraction. In the probe pad model, C_{ox}, models the oxide capacitance. The substrate is modeled by the parallel R–C network formed by R_{si} and C_{si}. The anodes of the varactor unit cells are connected with the top metal layer and the cathodes by the lower metal layer. The metal-to-bulk and metal-to-metal overlap capacitances are modeled by C_{ma}, C_{mk}, and C_{ol}. The base contact resistance is modeled by R_b. The actual varactor junction capacitance is C_j. The junction capacitance between the collector and substrate is C_{jb}. The resistance associated with the n region between the collector contacts and the active junction is R_c. In the oscillator circuit, the anode of the varactor is connected to the resonator and the cathode is at an RF ground. Figure 5.19 shows the measured and modeled capacitance seen at the anode of the varactor with the cathode at an AC ground. The measured and modeled (using the model of Figure 5.18) capacitance between the probe pads[4] is plotted in Figure 5.20. The effect of the interconnect inductance is clearly evi-

[4]Calculated by $C_{ka} = \frac{\Im y_{ka}}{2\pi f}$

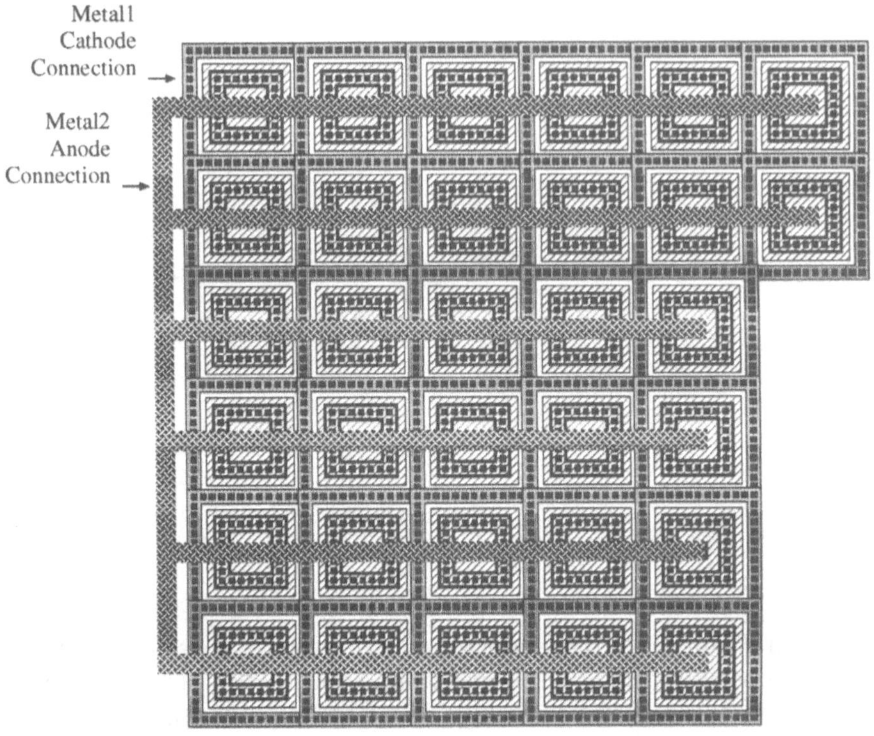

Figure 5.16. Varactor Layout

dent. Figures 5.21 and 5.22 show the measured and modeled reflection coefficients as the anode and cathode respectively of the test structure.

5.2.2 Bondwire Inductors

The bondwire inductors are formed by bonding between three on chip bond pads as shown in Figure 5.23. Low capacitance bond pads (i.e. the pads only include the top metal layer rather than all metal layers and gate polysilicon layer) reduce coupling into the lossy substrate. These inductors were characterized in [35] where it was shown that the inductance is approximately given by (5.1).

$$L_B(\text{nH}) \approx 0.13\text{nH} + 1.04\frac{\text{nH}}{\text{mm}}l \tag{5.1}$$

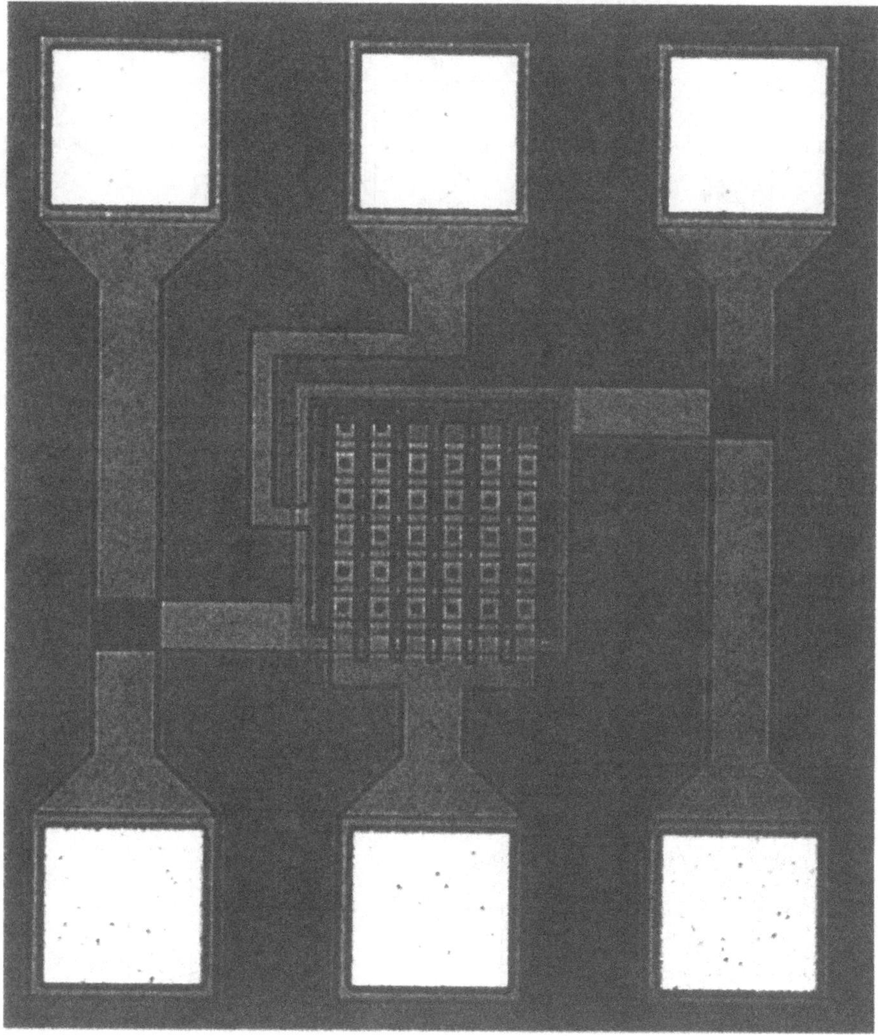

Figure 5.17. Varactor Test Structure Photo

The quality factor of the bondwire inductors at 1.8 GHz were reported to be around 12–13 in [35] and is approximately described by (5.2).

$$Q \approx 14.7 - \frac{1.02}{\mathrm{mm}} l \tag{5.2}$$

Figure 5.24 shows the circuit model for the inductors which was presented by [35].

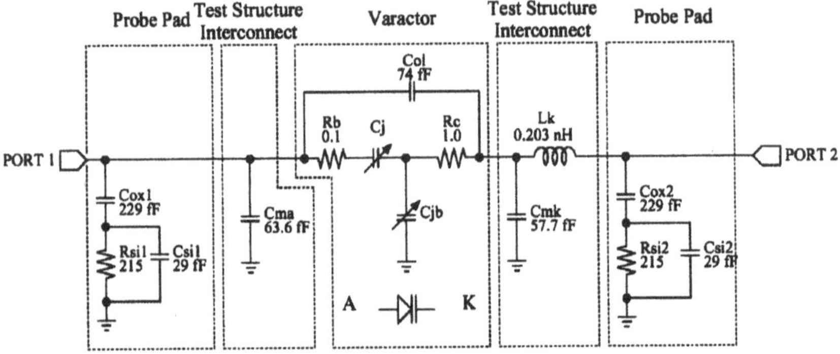

Figure 5.18. Model Used for Varactor Parameter Extraction

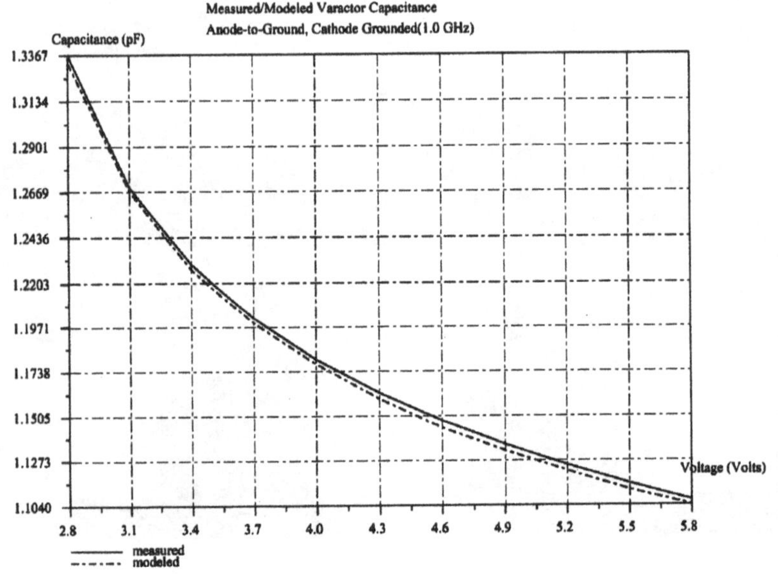

Figure 5.19. Varactor Measured C–V Characteristics

5.3. RF Phase Quantizer

This section describes the details of the integrated circuit realization of the RF phase quantizer circuit. The phase quantizer is based on the quadrature sampler approach discussed in section 5.1.5.

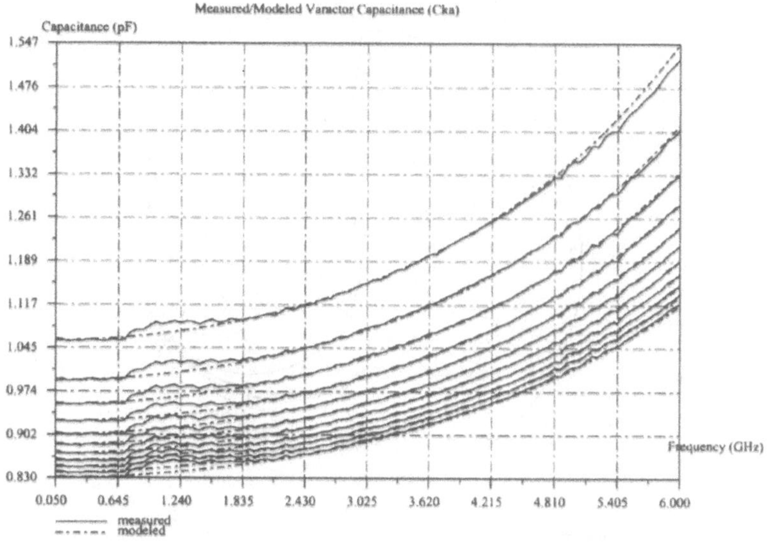

Figure 5.20. Varactor Measured Capacitance vs Frequency Characteristics

5.3.1 RF Phase Quantizer Circuit Details

Figure 5.25 shows the schematic of the phase splitter. The quadrature phase splitter is implemented as a fully differential RC–CR phase splitter. The differential input signal, $RF+$ and $RF-$, is buffered by Q1 and Q2. Resistors R1–4 in the quadrature network are base poly resistors. The quadrature network capacitors, C1–4, are poly–poly capacitors. The poly–poly capacitor structure is built on top of an N–well. This was done in an attempt to provide some isolation from the substrate. For the C–R section, capacitors C3 and C4 have their bottom plates driven by the buffer to avoid placing the additional bottom plate capacitance on the output nodes. Similarly, in the R–C section, capacitors C1 and C2 have their bottom plates on the ground side rather than the signal side. Resistors RB1 and RB2 provide the DC level required to bias the following stage.

The I and Q outputs of the quadrature network are sent to identical BiCMOS latches. Figure 5.26 shows the latch schematic. When the clock signal is low ($CLK-$ high and $CLK+$ low), Q1 and Q2 are biased on and the latch is transparent. When CLK goes high, the input differential

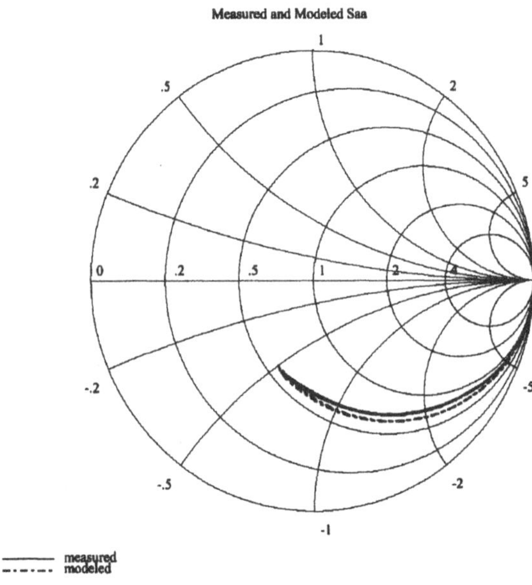

Figure 5.21. Varactor Measured S_{aa}

pair is turned off and Q3 and Q4 form a regenerative latch which quickly produces a full scale logic signal indicating the polarity of the RF signal at the time of the clock rising edge. The latch output is followed by an ECL to CMOS level converter to provide CMOS compatible logic swings. The use of NMOS transistors as the current switch allows a simple direct connection between the CMOS clock circuitry and the latch avoiding the need for explicit CMOS to ECL level translators. The use of CMOS devices in this fashion was inspired by a BiCMOS ripple adder circuit found in [36]. Figure 5.27 shows the schematic for the two phase clock driver circuit. As in [37], an overlapping clock scheme is used to reduce the voltage variation at the collector of the current source transistors in the latch. The use of overlapping clocks helps reduce coupling of the sample clock to the latch input. Figure 5.28 shows the ECL to CMOS level converter. This is a standard level shifter found in the literature, [38] and [39] for example.

Figure 5.29 shows the simulated differential input to the RF phase quantizer and the latch outputs when the latch is in transparent mode. The signals `v(inp)` and `v(inn)` are indicated in the quadrature network schematic in Figure 5.25. The signals `v(ip)`, `v(im)`, `v(qp)`, `v(qm)`,

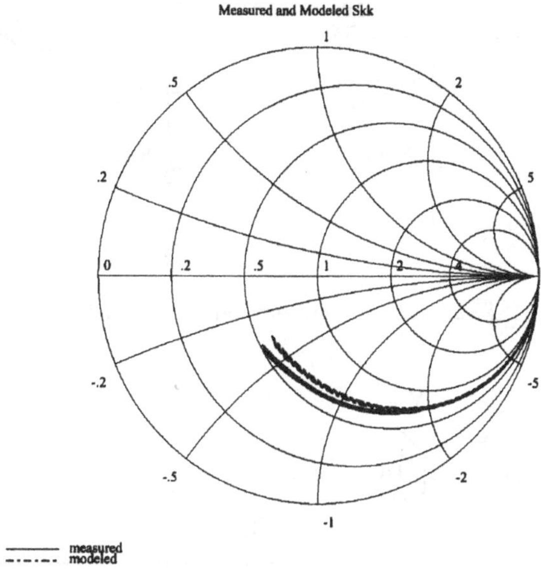

Figure 5.22. Varactor Measured S_{kk}

Figure 5.23. Bondwire Inductor Layout

`v(isamp)`, `v(qsamp)`, and `v(clk)` are indicated in the RF latch schematic in Figure 5.26. The quadrature relationship between the two outputs is clearly seen. Figure 5.30 shows the simulated operation of the RF phase quantizer over a sample clock cycle. The top trace is the clock signal. The next two traces are the differential outputs of the I and Q latches.

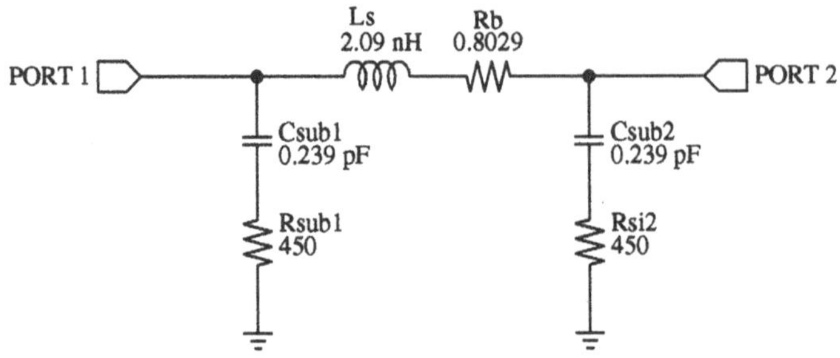

Figure 5.24. Bondwire Inductor Model

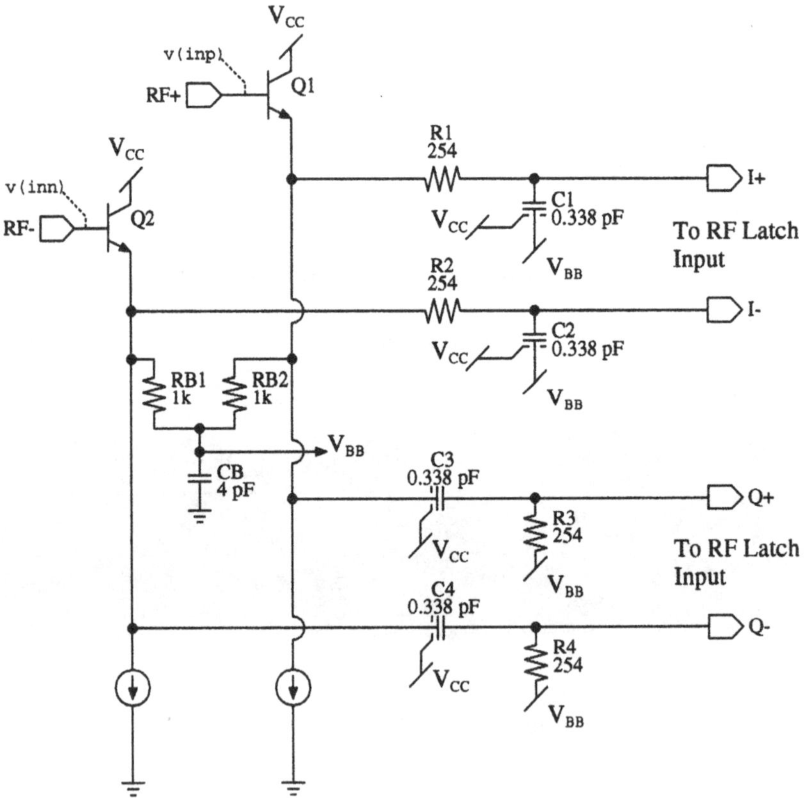

Figure 5.25. Integrated Circuit Quadrature Network

Finally, the lower two traces are the outputs of the ECL to CMOS level shifters. Figure 5.31 shows the simulated RF phase quantizer operation over several clock cycles.

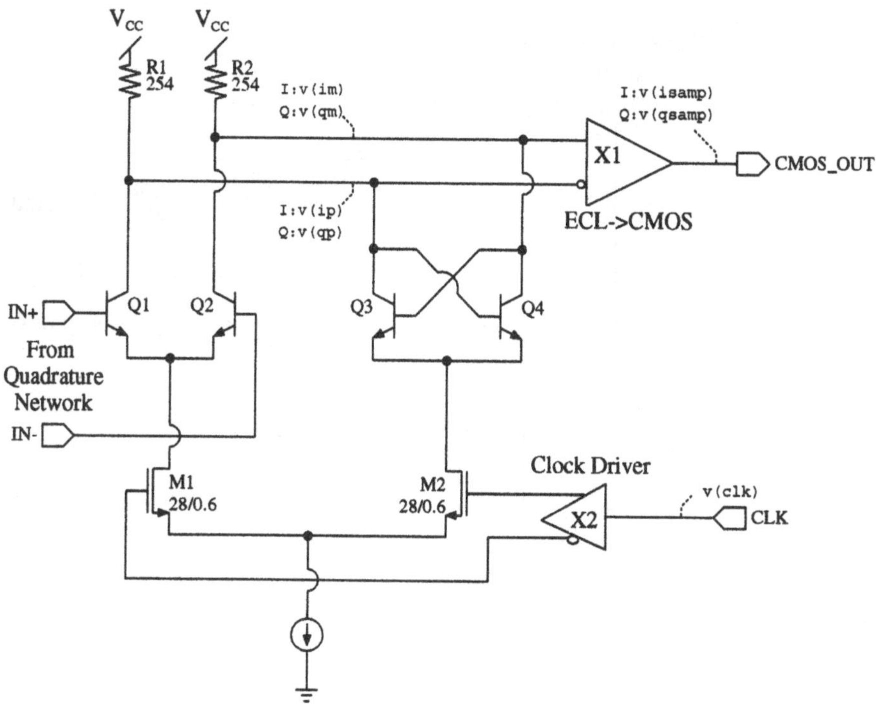

Figure 5.26. Integrated Circuit RF Latch

5.3.2 Phase Quantizer Comments

The implemented RF phase quantizer has reasonable performance and is a straightforward design. However there is room for substantial power savings. The input amplifier and regeneration amplifier in the RF sampler, Q1–Q4 in Figure 5.26, dissipate static power. Both phases of the sampler, amplification and regeneration/latching, are wasteful of power. During the amplification phase we only require the amplifier to reach steady state just in time for the clock rising edge. Examination of the waveforms in Figure 5.30 shows that it only takes a few nanoseconds for the amplifier to settle upon entering the amplification phase. In Figure 5.30 the amplification taking place from 7 ns to 30 ns represents wasted power. In addition, the regeneration/latching phase consumes static power. It should be possible to exploit these observations and only supply current to the latch during a small portion of the clock cycle. This approach has been successfully been applied to the design of clocked BiCMOS comparators [40].

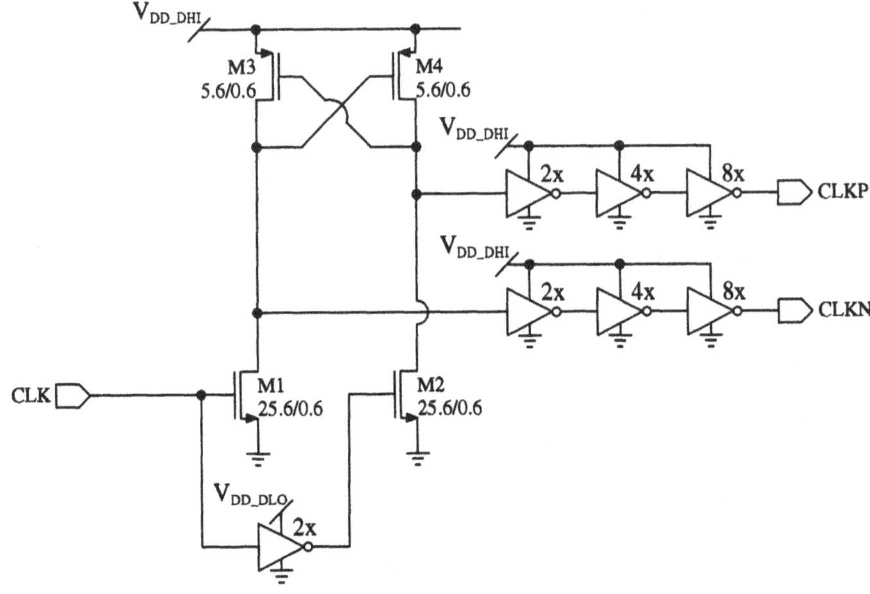

Figure 5.27. Two Phase Clock Generation Circuit

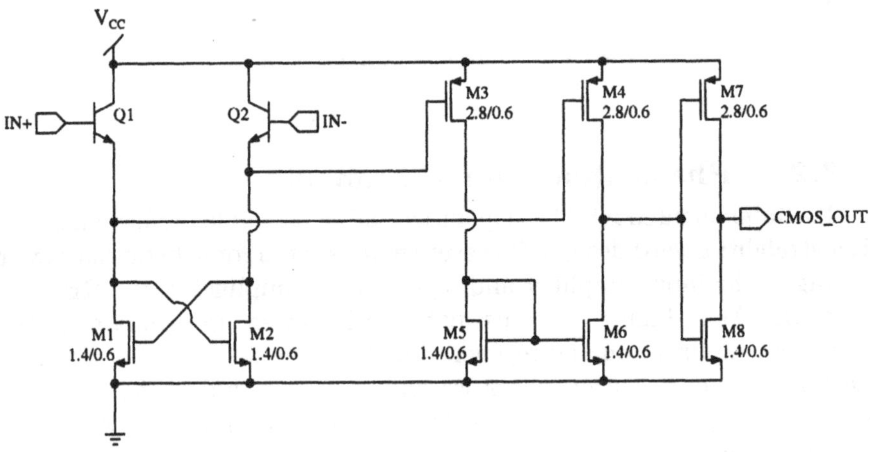

Figure 5.28. Integrated Circuit ECL to CMOS Level Converter

5.3.3 RF Interconnect

The interconnect in the test chip carrying the RF signals was designed to try and minimize its effect. Figure 5.32 shows a portion of the RF

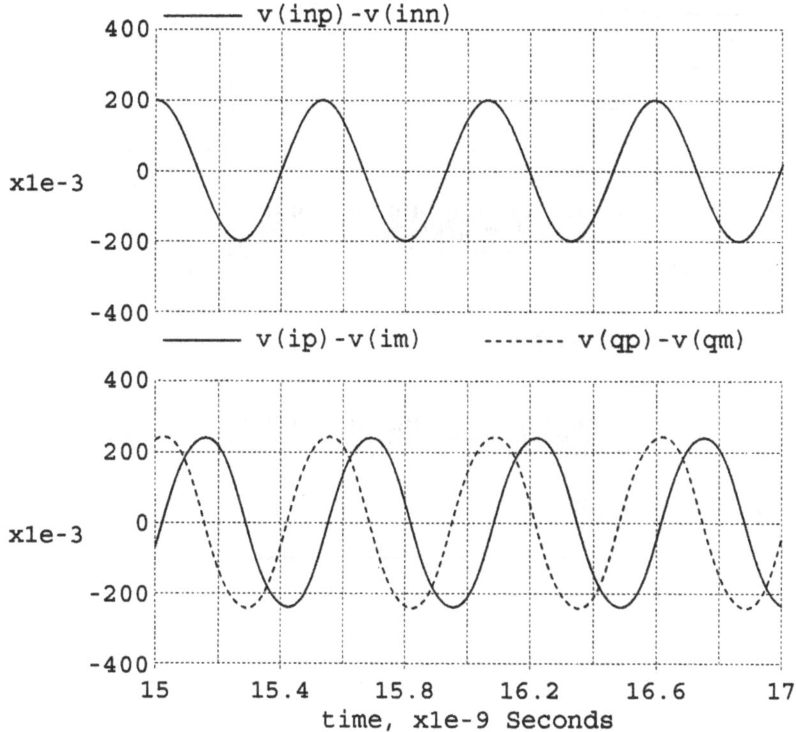

Figure 5.29. Integrated Circuit Quadrature Network Waveforms

interconnect used in the chip. The two coplanar waveguide traces at the lower right portion of the photo carry the signal from the VCO to the divider. The transistors in the center of Figure 5.32 make up the bipolar divide by two prescaler. The pair of coplanar waveguide traces at the upper right hand side of the photo connect the prescaler output to the CMOS divider input. At issue is the loss associated with the interconnect rather than the delay. The total line length is still significantly shorter than a wavelength. By using a coplanar waveguide structure, the fields are concentrated more in the gap between the metal trace and the metal ground rather than in the lossy substrate[41]. The traces are in the upper metal layer. Microstrip was not used because the oxide thickness between the two metal layers would dictate a very narrow line width to realize 50 ohms. The narrow line in turn would have high resistive losses.

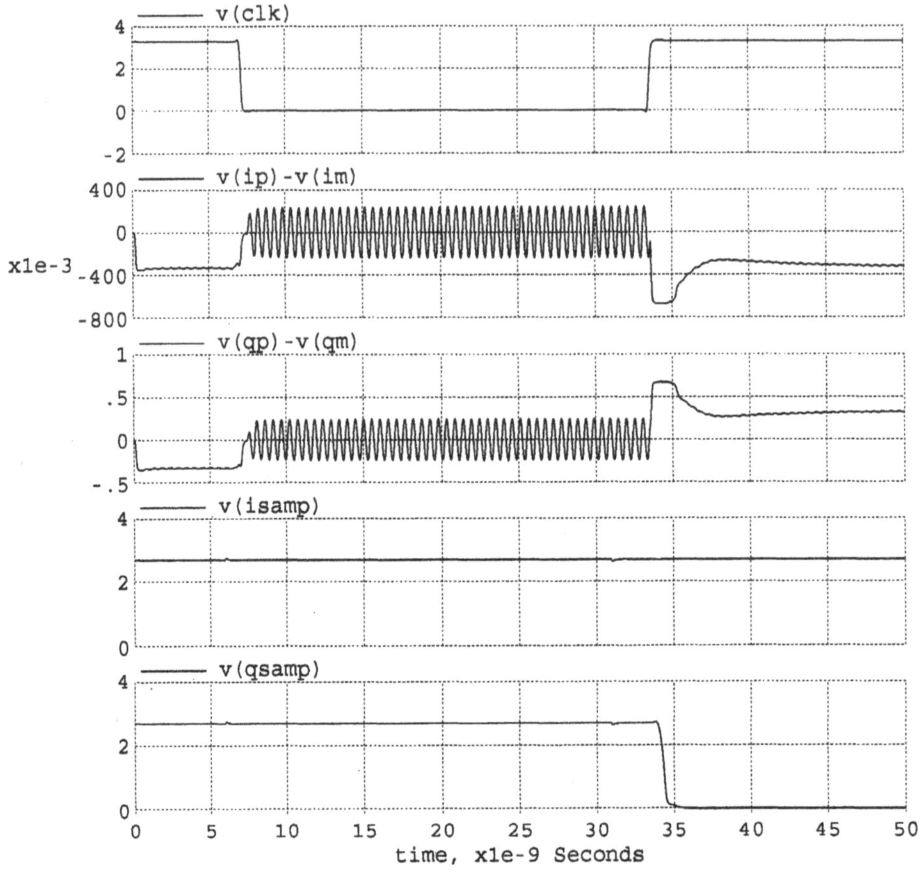

Figure 5.30. Integrated Circuit RF Phase Quantizer Waveforms (a)

5.4. Digital Building Blocks

This section describes the main building blocks used in the digital signal processing circuitry. All of the internal digital logic runs from a low supply voltage of nominally 1.5 Volts.

5.4.1 Flip–Flops

There are two types of flip–flops used in the digital section. The fully dynamic flip–flop shown in Figure 5.33 is used in all of the signal processing circuits. A fully static version of the flip–flop, shown in Figure 5.34, is used for the configuration registers. The use of local clock inversion in each flip–flop simplifies the clock routing and limits the amount of skew between the two clock phases.

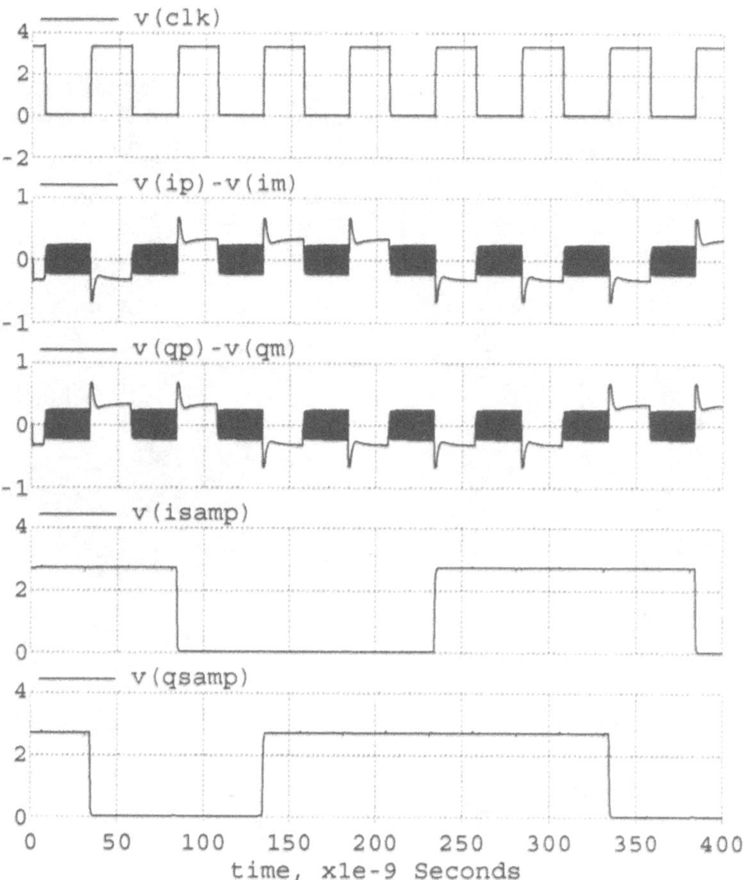

Figure 5.31. Integrated Circuit RF Phase Quantizer Waveforms (b)

5.4.2 Adders

The adder cells used in the calibration circuit are shown in Figure 5.35 and Figure 5.36. The two adder cells differ in the sign of the carry in/out signals. Multi-bit adders are formed by using Figure 5.35 for the odd numbered bits and Figure 5.36 for the even numbered bits. The logic gates are minimum sized static CMOS style gates.

5.5. Reference Error Filter

In section 4.2, it was shown that the error between the RF output phase and desired output phase is approximately equal to the gain error multiplied by the reference error waveform, $\phi_{e,ref}$. The reference error

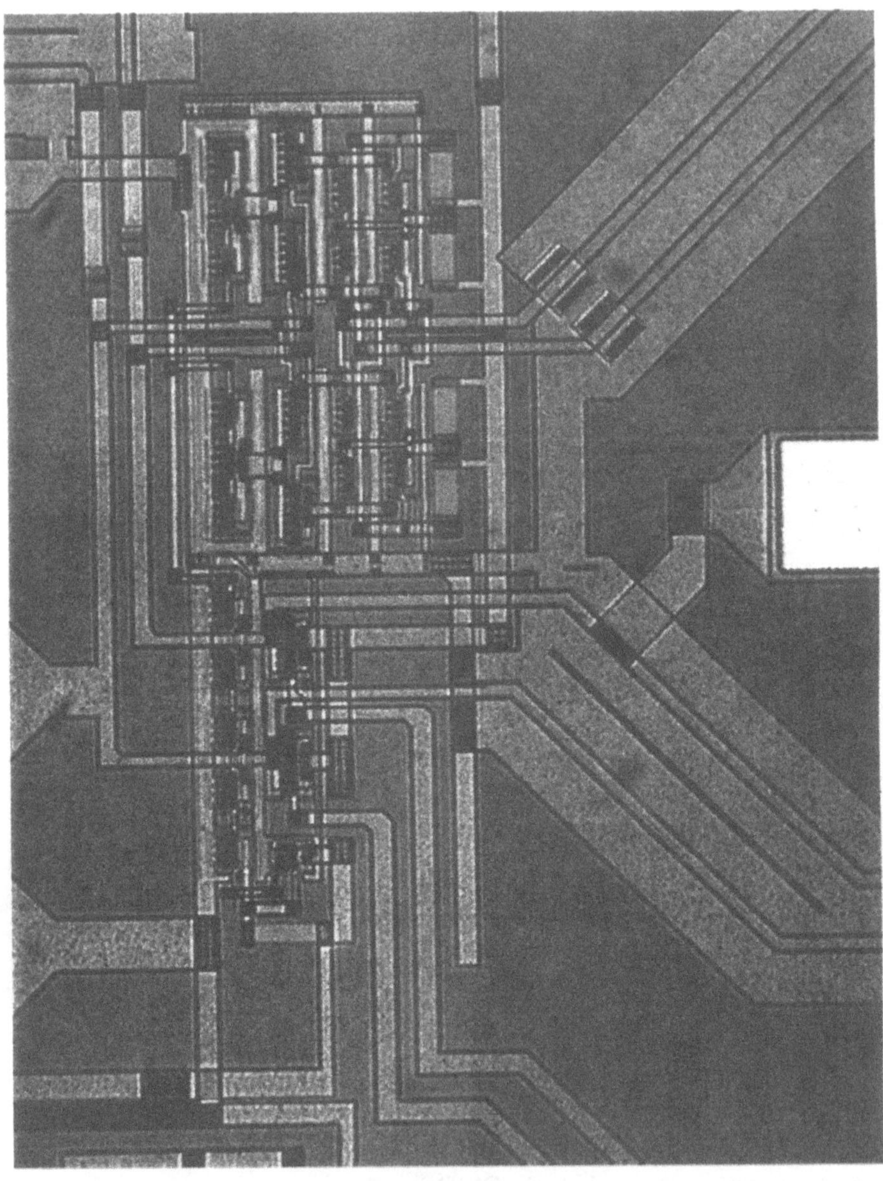

Figure 5.32. Bipolar Prescaler and RF Interconnect Photo

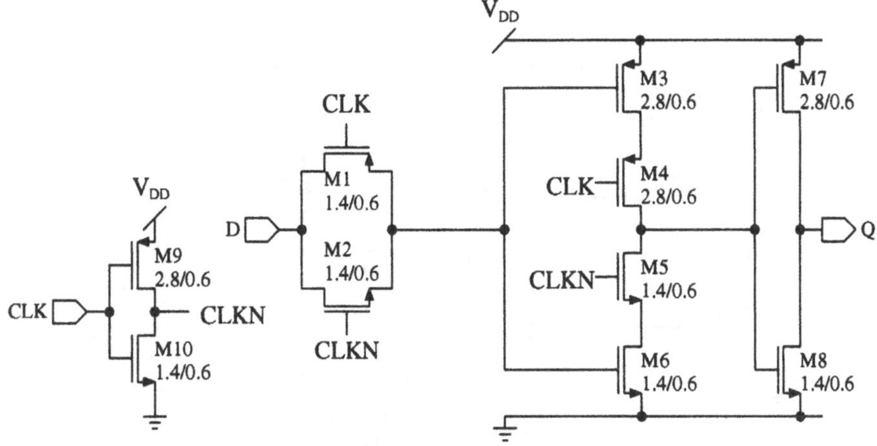

Figure 5.33. Dynamic CMOS Flip–Flop

was shown to be related to the ideal modulation waveform by the transfer function given in (5.3).

$$\Phi_{e,ref}(s) = \left(\frac{s}{K_{T_0}}\right)\left(\frac{\tau_{p_0} s + 1}{\frac{\tau_{p_0}}{K_{T_0}} s^3 + \frac{1}{K_{T_0}} s^2 + \tau_{z_0} s + 1}\right)\Omega_{mod}(s) \qquad (5.3)$$

Note that the scale factor, $\left(\frac{1}{K_{T_0}}\right)$, in front of (5.3) is not important and can be set to whatever value is convenient. The job of the reference error filter is to approximate the transfer function in (5.3). The nominal PLL loop bandwidth is around 80 kHz and the system clock frequency is 20 MHz. This suggests that implementing the complete error filter with a full 20 MHz clock rate will be somewhat wasteful of power [42]. If an infinite impulse response (IIR) filter is used, the adders will have to be fast enough to not require pipelining[5]. This would lead to an increased power supply requirement and associated power consumption increase. If a finite impulse response (FIR) filter is used, the relatively narrow[6] transition band will lead to a large number of filter coefficients. These considerations indicate that the reference error filter should not be run at the full clock rate.

[5]The latency associated with pipelining is not compatible with the feedback loops associated with IIR filters.

[6]Narrow relative to the sample rate.

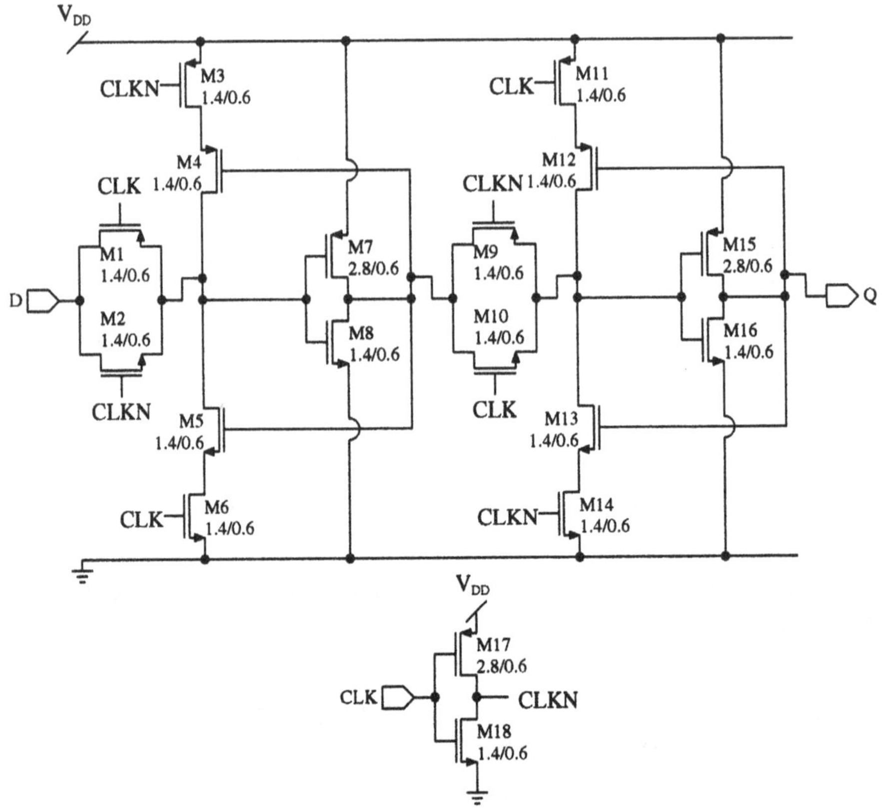

Figure 5.34. Static CMOS Flip-Flop

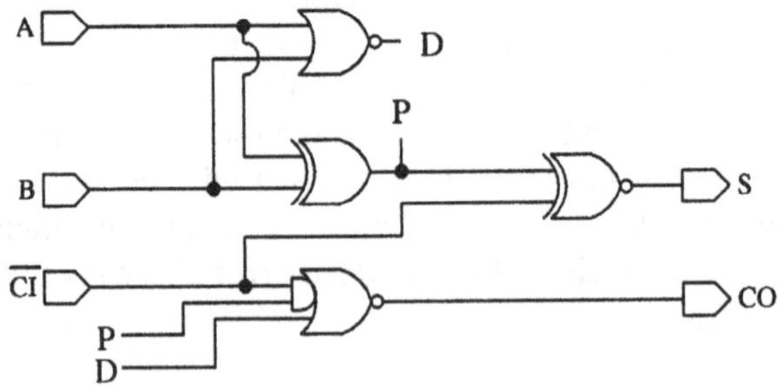

Figure 5.35. Odd Stage CMOS Full Adder

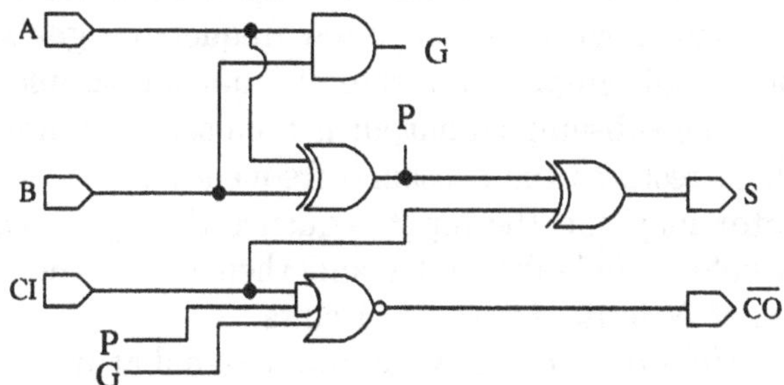

Figure 5.36. Even Stage CMOS Full Adder

The implemented reference error filter consists of a decimation stage [43] which reduces the sample rate by a factor of 32. The decimator output then feeds an FIR approximation to the desired filter response. Figure 5.37 shows the reference filter block diagram.

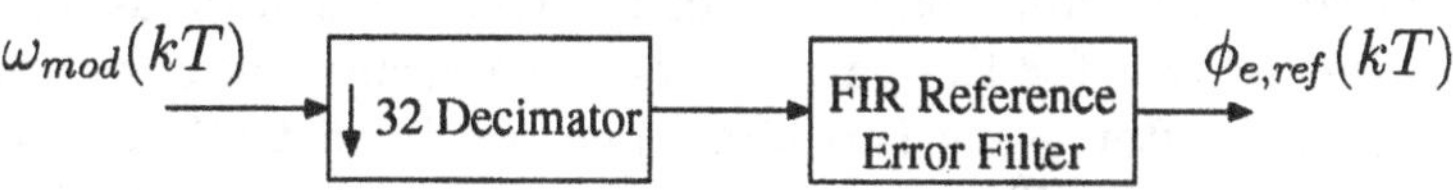

Figure 5.37. Reference Error Filter Block Diagram

5.5.1 Decimator Design

The decimator used by the reference error filter is the cascaded integrator comb (CIC) type. This is also known as a Hogenauer [44] filter. Figure 5.38 shows the block diagram for a CIC decimator. In Figure 5.38,

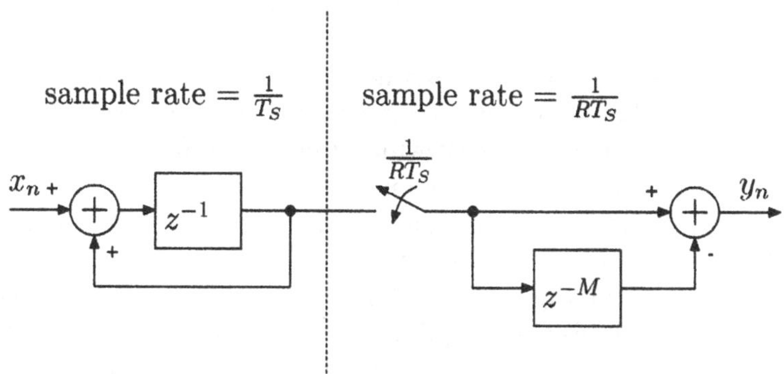

Figure 5.38. CIC Decimator

the input sequence, x_n, is accumulated. This reduces the amplitude of high frequency signals with respect to low frequency signals. The accumulator output is subsampled at a rate of 1 output sample for every R input samples. The subsampled output is then sent to a finite differencing circuit which restores the passband frequency response. In general a CIC decimator may run the input sequence through a cascade of N accumulators before the subsampler and then follow the subsampling with a cascade of N finite difference circuits.

For a CIC decimator of order N, an input signal at frequency ω radians/second will have its power scaled by the factor in (5.4).

$$H_{CIC}(\omega) = \left[\frac{\sin\left(\frac{RM\omega}{2}\right)}{\sin\left(\frac{\omega}{2}\right)}\right]^{2N} \tag{5.4}$$

Of course frequencies above half of the output sample rate will be aliased. In this case, (5.4) gives the scale factor for the aliased signal relative to a DC input.

The integrated circuit realization of the CIC decimator uses $N = 1$, $R = 32$, and $M = 1$ for the parameters. The accumulator is pipelined using a register after every two bits. The pipelined CIC decimator is shown in Figure 5.39. The use of pipelining allows the use of a low sup-

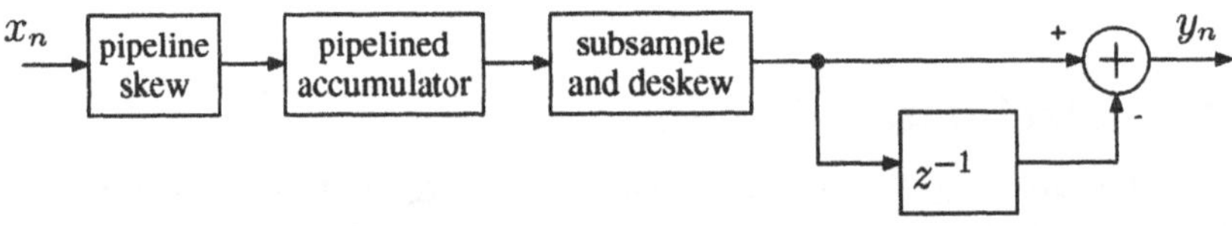

Figure 5.39. Pipelined CIC Decimator

ply voltage and results in reduced power consumption. We can use the fact that most (31/32) of the accumulator output samples are discarded to simplify the design of the deskewing circuit required to terminate the pipeline. To deskew the output of the pipelined accumulator, we simply sample the different accumulator output bits on different clock cycles of the master clock. The generation of the active low clock enable pulses is shown in Figure 5.40. The falling edge of the low rate clock, *CLKDEC*, initiates the sampling of the accumulator output. The clock enable signals, CEN_{0-5}, are used to enable the registers shown in Figure 5.41. Since the outputs of the registers feed into circuitry which is clocked from the rising edge of *CLKDEC*, the different sample times

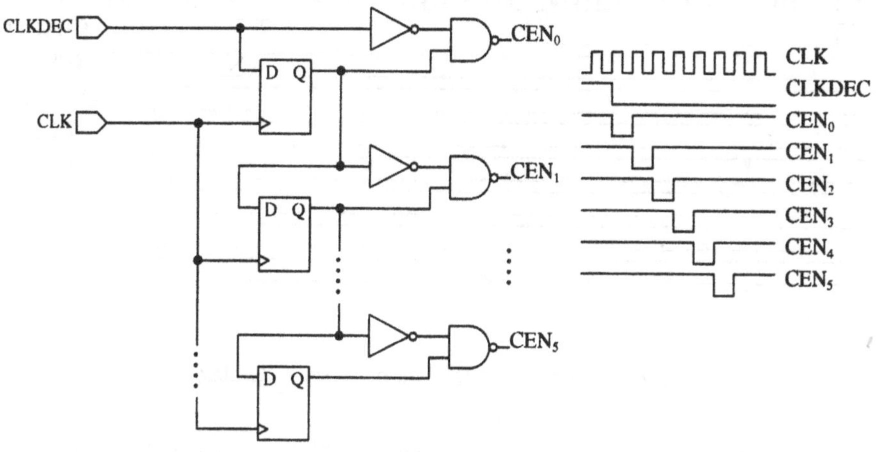

Figure 5.40. CIC Decimator Deskew Logic

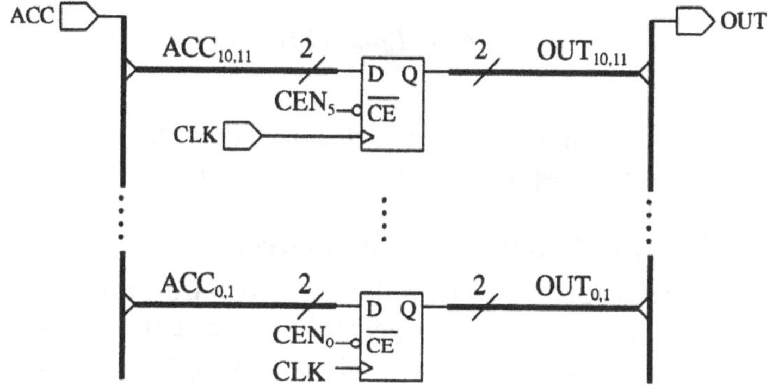

Figure 5.41. CIC Decimator Deskew Registers

do not cause any skewing of the data. By performing the deskewing in this fashion, we save using the 30 additional registers which would be required by deskewing the accumulator output in the usual way. The operation of the combined subsampling and deskewing part of the CIC decimator is illustrated in Figure 5.42.

5.5.2 Discrete Time Filter Approximation

An FIR type filter is used for the reference error filter due to the simple hardware implementation which is possible. The reference error filter output needs to match the time domain response implied by (5.3).

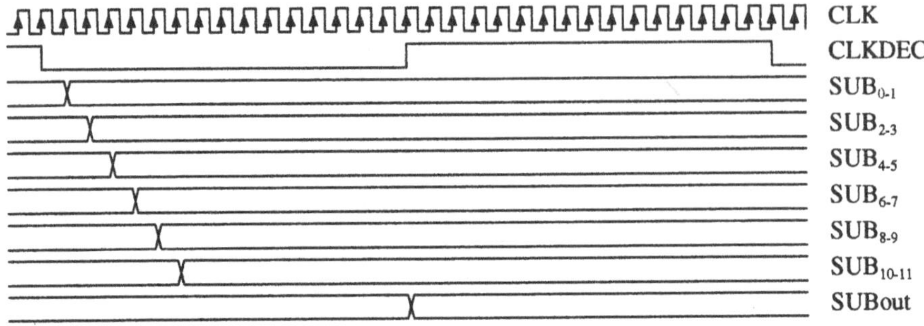

Figure 5.42. CIC Decimator Deskew Timing

As such, the impulse invariance method [45] was used to produce the ideal discrete time impulse response from the continuous time transfer function in (5.3). In this method, the discrete time impulse response is obtained by sampling the continuous time impulse response as illustrated by (5.5).

$$h_d[n] = T_d h_c(nT_d) \tag{5.5}$$

The ideal continuous time impulse response and the ideal truncated discrete time impulse response are shown in Figure 5.43.

5.5.3 Digital Filter Architectures

There are several options available for the hardware implementation of the FIR filter. Figure 5.44 shows a general purpose FIR filter structure. The output sequence, y_n, is the result of convolving the input sequence, x_n, with the filter impulse response, b_n.

The implementation of the multipliers in Figure 5.44 has a big influence on the power consumption and physical size of the filter. In situations like the current problem where the process can support a higher clock rate than the sample clock, the multipliers can be simply implemented as serial multipliers. A serial multiplier is shown in Figure 5.45. In Figure 5.45 the input signal, x, is loaded into a shift register. As the data is shifted out one bit at a time, the coefficient, b, is multiplied by the single bit signal. The results of the multiplications are scaled and accumulated to produce the final result. This approach requires M clock cycles to process an M-bit input signal. With this type of multiplier, the space of allowed coefficients is the set of all N-bit numbers. From a power consumption point of view, there are some disadvantages to this approach. We require a clock which runs at M times the filter sam-

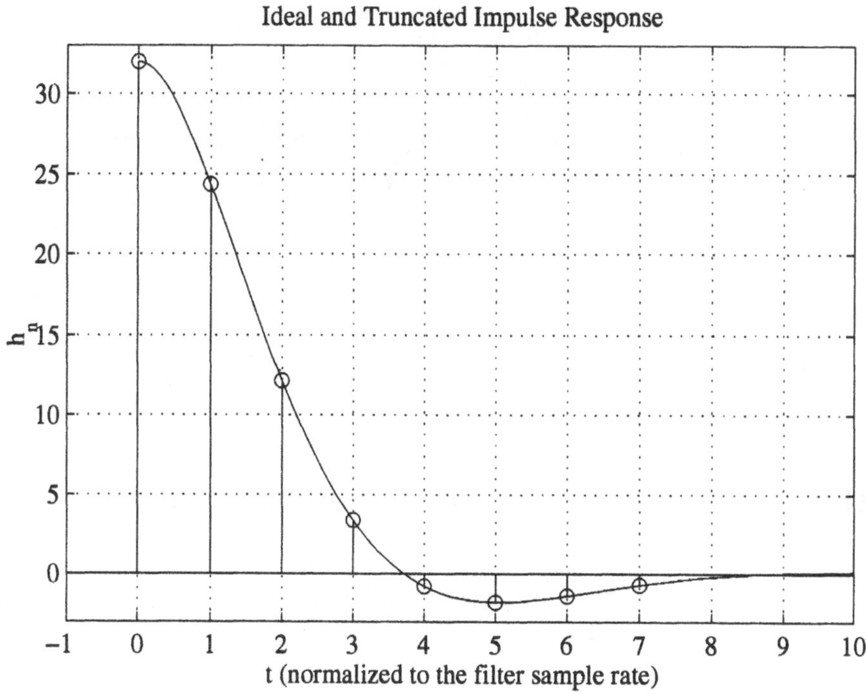

Figure 5.43. Error Filter Ideal Impulse Response

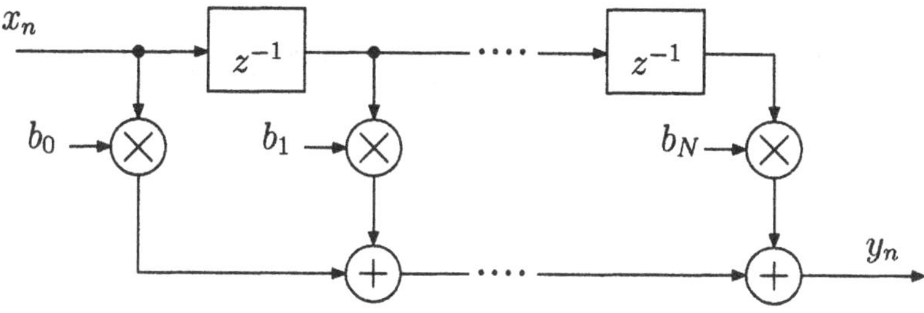

Figure 5.44. General Purpose Direct Form FIR Filter Structure

ple rate. The accumulator must operate at the higher frequency which can in turn drive up the power supply voltage requirements. The other drawback is that even when the b coefficient contains a large number of zeros in it, we still perform the same amount of processing.

A second approach which requires M clock cycles to process an M-bit input signal is the distributed arithmetic [46] filter shown in Figure 5.46.

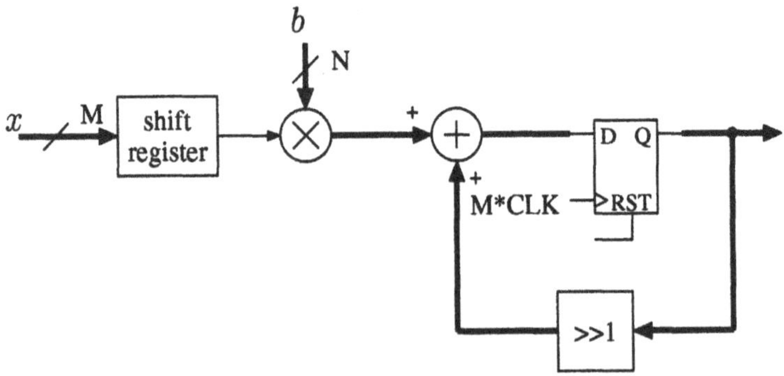

Figure 5.45. Serial Multiplier

The distributed arithmetic filter works on the following principle. The

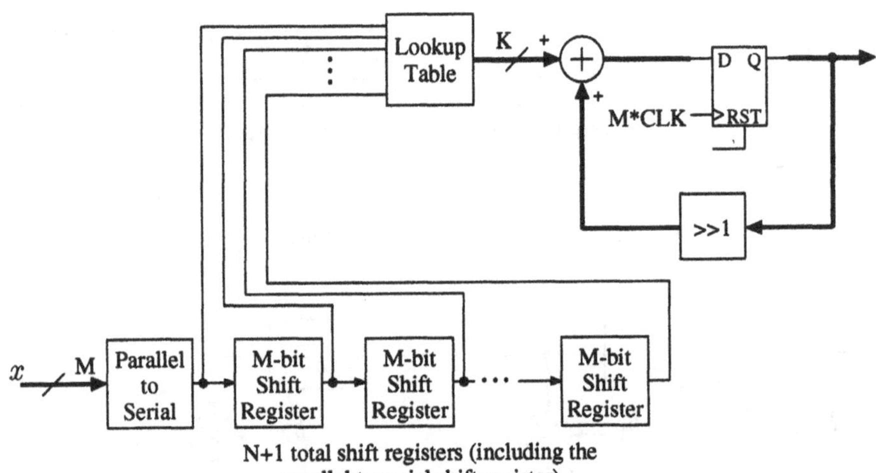

Figure 5.46. FIR Filter Based on Distributed Arithmetic

filter output sequence is given by the convolution sum (5.6).

$$y_n = \sum_{j=0}^{N} b_j x_{n-j} \tag{5.6}$$

The sum can be expanded by explicitly representing the input samples as a binary number. In (5.7), the notation $x_{n,b}$ refers to bit b of sample

n of the signal x.

$$y_n = \sum_{j=0}^{N} b_j \left(\sum_{k=0}^{M-1} x_{n-j,M-1-k} 2^{M-1-k} \right) \tag{5.7}$$

Now we exchange the order of the summations in (5.7) to obtain (5.8).

$$y_n = \sum_{k=0}^{M-1} \left(\sum_{j=0}^{N} b_j x_{n-j,M-1-k} \right) 2^{M-1-k} \tag{5.8}$$

Now note that for a given set of filter coefficients, the inside sum, $\sum_{j=0}^{N} b_j x_{n-j,M-1-k}$, can only take on one of 2^{N+1} possible values. These 2^{N+1} values can be stored in a RAM or ROM lookup table. This function is the purpose of the lookup table in Figure 5.46. The shift registers, accumulator, and shifter explicitly implement the outer sum in (5.8). To implement a filter of length $N + 1$ with M–bits of precision on the input samples and L–bits on the output samples we require a $2^{N+1} \times L$ RAM/ROM that operates quickly enough to support clocking the accumulator at M times the filter sample rate. In cases where the filter length causes the lookup table to become too large, the sum in (5.8) may be subdivided and the resulting partial sums may be combined later. This operation can be pipelined if required. When the required sample rate becomes to high to allow a clock running at M times the sample rate, the upper bits may be processed in parallel with the lower bits. This gives a direct tradeoff between die area and clock speed. Processing the bits in parallel can also be used to reduce power consumption [47]. For example, with the filter sample rate and supply voltage held constant, power consumption is largely unchanged when we split the circuit computation into one circuit for the MSB's and one for the LSB's[7]. Since we have doubled the circuit size, the switched capacitance doubles but the clock frequency is halved which keeps the same power. However, we can now lower the supply voltage due to the slower speed requirement and thus the power is reduced. Despite its attractive features, the distributed arithmetic filter still does not exploit the zero valued digits in the coefficients.

An alternative approach is to implement the multipliers in a fully parallel fashion. By expressing the coefficients as a sum of signed powers of two (SPT), the multipliers can be implemented using adders and without requiring a higher speed clock. For example, consider the task

[7]Actually, the power increases slightly since we need one additional adder to combine the two outputs.

of multiplying by 27. We can express this as $27 = 32 - 4 - 1 = 2^5 - 2^2 - 2^0$. Figure 5.47 shows the implementation of a multiply by 27. In the literature this representation is often times referred to as "canonic signed digit" or CSD. There is a simple set of rules which the canonic signed digit representation must satisfy [48]. The rules are that the canonic representation must have the minimum number of non–zero digits and no two adjacent digits may be non–zero. For example $3 = 2^1 + 2^0$ satisfies the minimum number of non–zero digits constraint but is not the canonic form because the two non–zero digits are adjacent. The canonic representation is $3 = 2^2 - 2^0$. Since we may choose to actually implement a multiply by 3 using the first expansion we will simply refer to these multipliers as SPT rather than CSD. The signed powers of two multiplier

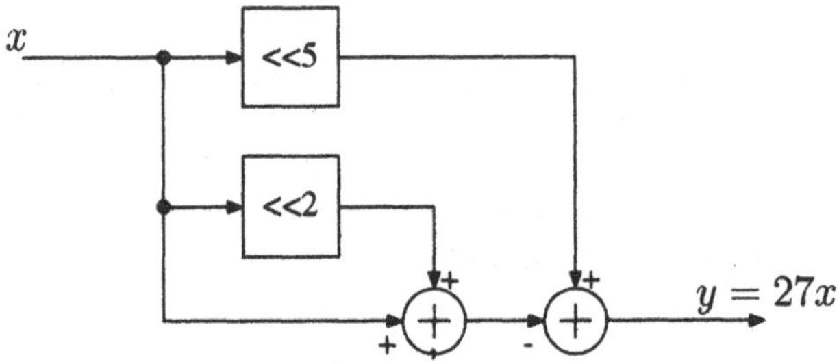

Figure 5.47. Signed Powers of Two Multiplier

is attractive for a couple of reasons. If the filter coefficients contain a relatively small number of non–zero terms, the resulting structure can be fairly simple. The other advantage is that the adders all operate at the filter sample rate instead of M times the sample rate as with the serial multiplier. Due to its parallel nature the signed powers of two approach tends to be somewhat more area intensive than serial approaches. For this project, however, the filter area is not too large and as such the signed powers of two implementation was used.

5.5.4 SPT Approximation

After obtaining the ideal discrete time impulse response, the coefficients must be quantized to allow implementation in digital hardware. To make effective use of the signed powers of two structure, we need to limit the number of non–zero digits in each coefficient. This leads to an interesting coefficient optimization problem. To illustrate the problem, consider the set of all 7–digit numbers that have no more than 2 non–zero digits. The value of each digit is ± 1 or 0. This leads to the quantization curve shown in Figure 5.48. As noted in [49], the error

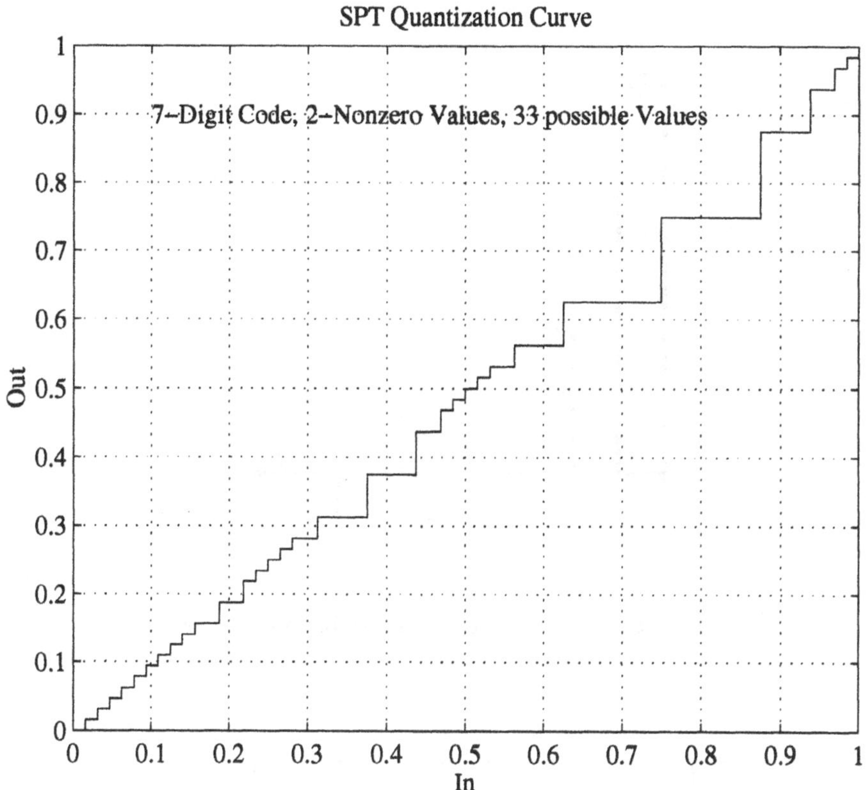

Figure 5.48. 7–Digit, 2 Non–Zero Digit SPT Quantization Curve

associated with quantizing a set of filter coefficients can be reduced by applying a fixed scale factor to all of the coefficients prior to quantizing. The optimal scale factor causes most of the coefficients to end up in the denser regions of the SPT space.

A modified version of the optimization algorithm given in [49] was applied to yield the final reference error filter impulse response shown in Figure 5.49. The cost function was a normalized squared error in the impulse response rather than the maximum frequency domain error cost function used in [49]. The reference error filter was implemented in the transposed form [45] shown in Figure 5.50. The transposed form avoids the problem of having to add all $N + 1$ taps in one clock cycle as in the direct form filter in Figure 5.44. The disadvantage of the transposed form is the large capacitive input load.

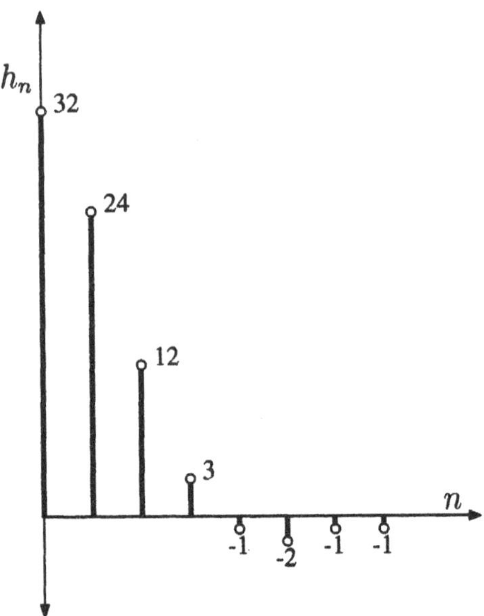

Figure 5.49. Final Reference Filter Impulse Response

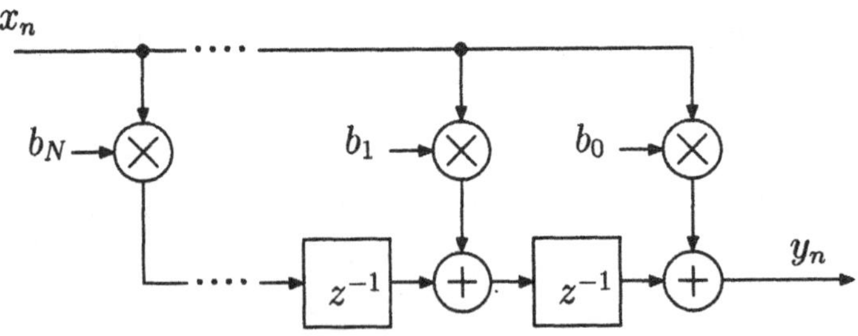

Figure 5.50. Transposed Form FIR Filter

Examination of the SPT representation of the filter response reveals some redundancy that can be exploited to simplify the filter structure [50]. Table 5.2 lists the filter coefficients and their SPT representations. We see that h_1, h_2, and h_3, all share the term $(2^1 + 2^0)$. We only need to calculate this once and the result is shared by the other two taps. Figure 5.51 shows the schematic for the implemented reference error FIR filter. Note that the minus sign associated with b_{4-7} is achieved by using a subtractor in place of the fourth adder. This eliminates the need for using several two's complementers.

n	h_n	*SPT Code*
0	32	2^5
1	24	$2^3\left(2^1+2^0\right)$
2	12	$2^2\left(2^1+2^0\right)$
3	3	2^1+2^0
4	-1	-2^0
5	-2	-2^1
6	-1	-2^0
7	-1	-2^0

Table 5.2. Reference Error Filter SPT Coefficients

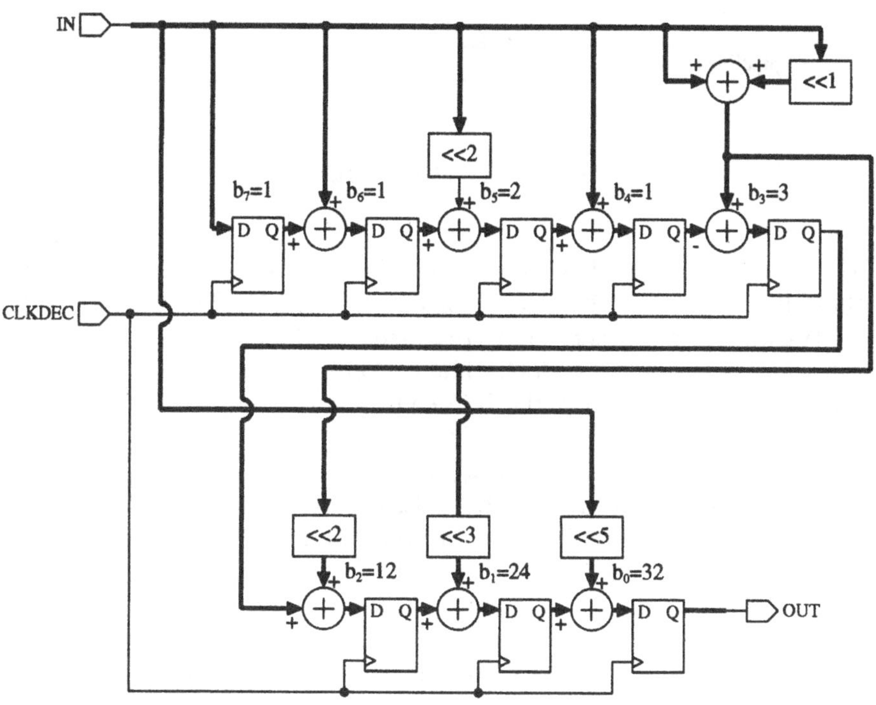

Figure 5.51. Reference Error Filter Schematic

5.6. Digital Phase Locked Loop and Phase Error Measurement

Figure 5.52 shows the top level schematic for the phase error measurement circuit and digital phase locked loop. The sampled and quantized

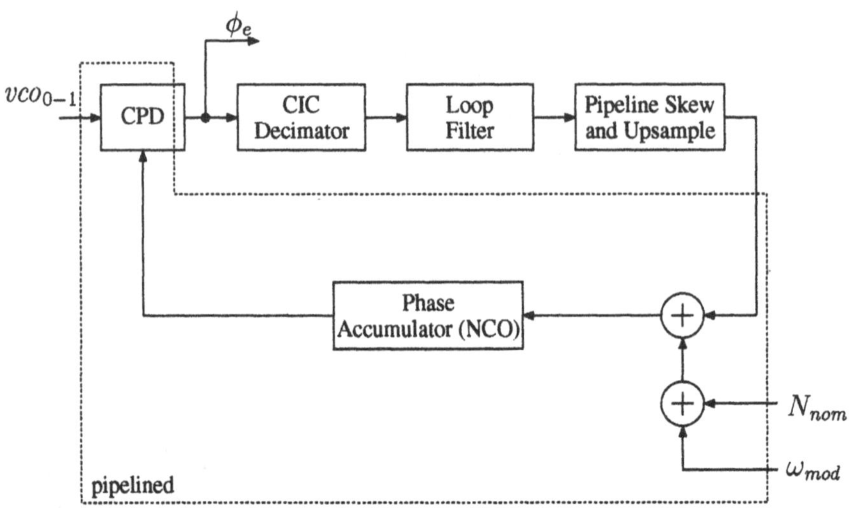

Figure 5.52. DPLL Schematic

RF phase is compared to the numerically controlled oscillator (NCO) phase by the carrier phase detector (CPD) circuit. The CPD includes the track and hold functionality discussed in Chapter 4. The CPD output is the output of this block. The CPD output also drives the DPLL loop filter which controls the NCO phase to remove any static phase offset and drift.

The nominal divide value N_{nom}, which sets the NCO center frequency is added to the loop filter output. In addition, the desired frequency modulation signal is added in. Due to scaling issues which will be discussed later, the loop filter is operated at a lower clock frequency than the NCO and CPD. To support operation at the low supply voltage, pipelining is used in the higher clock rate circuits. Since N_{nom} is a static value, its bits do not require skewing before entering the pipeline. Since ω_{mod} also drives the pipelined reference error filter, the skewing circuit can be shared.

5.6.1 DPLL Loop Dynamics

For the purposes of determining the bandwidth and stability of the digital phase locked loop, we use the simplified block diagram in Figure 5.53. In the block diagram, K_D represents the phase detector gain, K_{CIC} is the *DC* gain of the decimator, K_P and K_I are the proportional path and integral path gain of the loop filter and $K_{\uparrow}$ represents the gain

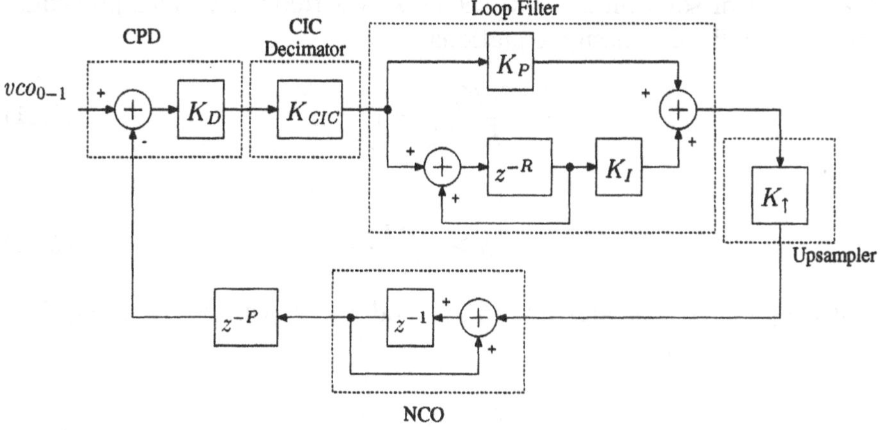

Figure 5.53. Block Diagram For DPLL Analysis

associated with the process of upsampling the loop filter output. The delay element in the loop filter accumulator is z^{-R} rather than z^{-1} because the accumulator is clocked at a rate which is a factor of R lower than the main system clock. The delay term z^{-P} accounts for the additional delay around the loop due to the pipelining. Note that we are ignoring the dynamics associated with the decimator. Since the loop bandwidth will be much less than the lower sample rate, this omission is justified.

The phase detector gain is unity, $K_D = 1$. The CIC decimator has also been designed for unity gain, $K_{CIC} = 1$. The loop filter output is upsampled by zero stuffing its output. This results in a gain of $K_\uparrow = \frac{1}{R}$. At this point, we can make two more simplifying assumptions. Since the loop bandwidth will be a small fraction of the lower sample rate, we will ignore the z^{-P} term and approximate the accumulator frequency response as (5.9).

$$\frac{z^{-1}}{1 - z^{-1}} \Big|_{z=e^{j\omega T_s}} \approx \frac{1}{j\omega T_s} \tag{5.9}$$

Under these assumptions the $DPLL$ loop gain is given by (5.10).

$$L(j\omega) = \left(K_P + \frac{K_I}{j\omega R T_s}\right)\left(\frac{1}{R}\right)\left(\frac{1}{j\omega T_s}\right) \tag{5.10}$$

A simple design approach is to set $K_I = 0$ and choose R and K_P to achieve the desired crossover frequency. Next choose K_I to place the zero at or somewhat below the crossover frequency. This procedure yields the following design equations.

$$\frac{K_P}{R\omega_c T_s} = 1 \tag{5.11}$$

$$K_P \geq \frac{K_I}{\omega_c R T_s} \tag{5.12}$$

With a little bit of algebra, (5.11) and (5.12) can be rearranged to give (5.13) and (5.14).

$$K_P = \omega_c T_s R \tag{5.13}$$

$$K_I \leq K_P^2 = (\omega_c T_s R)^2 \tag{5.14}$$

The reason for the decimator will now become clear. Suppose the decimator was omitted ($R = 1$). The sample rate is 20 MHz and the desired loop bandwidth is around 1 kHz. From (5.13) and (5.14) we obtain $K_P = 2^{-11.6}$ and $K_I = 2^{-23.3}$. These very small gain terms present some implementation difficulties. To realize the proportional gain, we must have at least a few more bits than 12 (the number which will be lost) coming from the CPD. Similarly, the accumulator would need to have enough bits to obtain the desired tuning range even after discarding 24 bits. Also note that in this case, the accumulator is clocked at the full clock rate. For these reasons, the loop filter is operated at a lower clock frequency which reduces the required bit width and also the required speed of the loop filter clock components.

In the implementation of the DPLL, the ratio, R, of the master clock frequency to the loop filter clock frequency is set to be 32. Now when we apply (5.13) and (5.14) we obtain $K_P = 2^{-6.6}$ and $K_I = 2^{-13.3}$. These values are easier to implement. Also note that the accumulator now runs at $\frac{20\text{MHz}}{32} = 625\text{kHz}$ which allows the use of a low supply voltage without requiring pipelining.

5.6.2 Carrier Phase Detector with Track and Hold

The carrier phase detector accepts the samples from the RF phase quantizer and the NCO. The phase difference is computed and sent to a track and hold circuit. The incoming values from the RF phase quantizer are gray coded and must be converted to a two's complement representation. In addition, the VCO samples must be skewed before driving the pipelined subtractor. Figure 5.54 shows the schematic of the complete phase difference circuit. Figure 5.55 shows the schematic of the

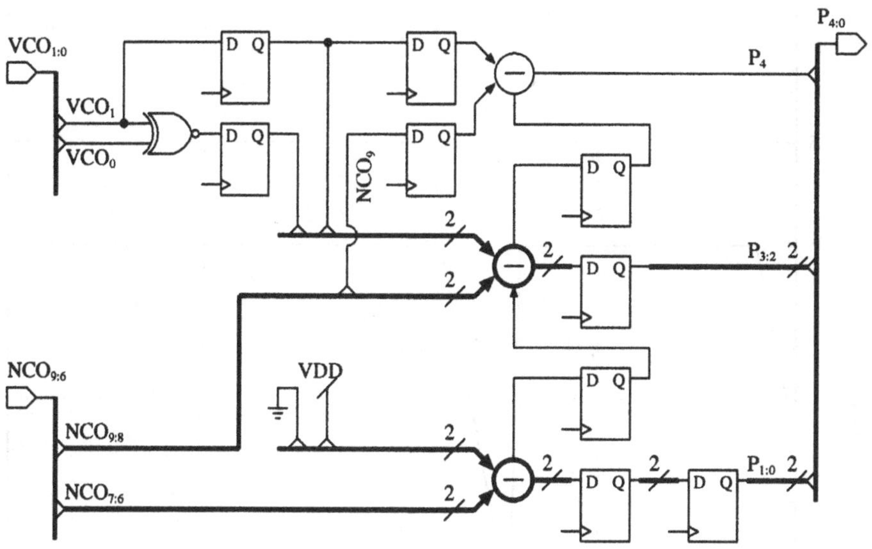

Figure 5.54. DPLL Phase Difference Circuit

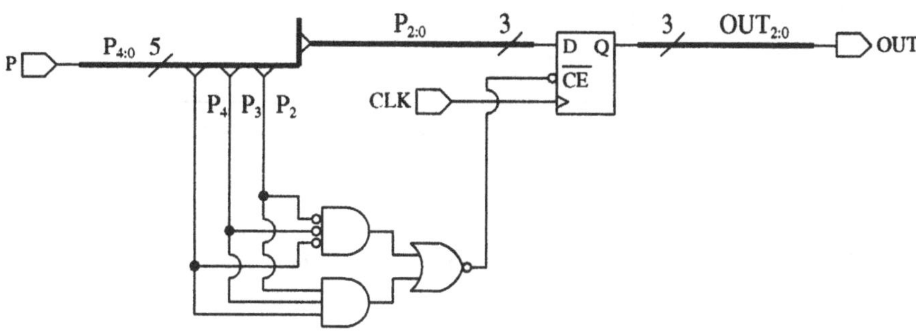

Figure 5.55. Track and Hold Implementation

track and hold implementation. When the three most significant bits of the input are identical, the output register is enabled and the input is passed to the output. Otherwise the previous output is held. Figure 5.56 shows the simulated output of the track/hold circuit with a positive going phase ramp at its input. Figure 5.57 shows the track and

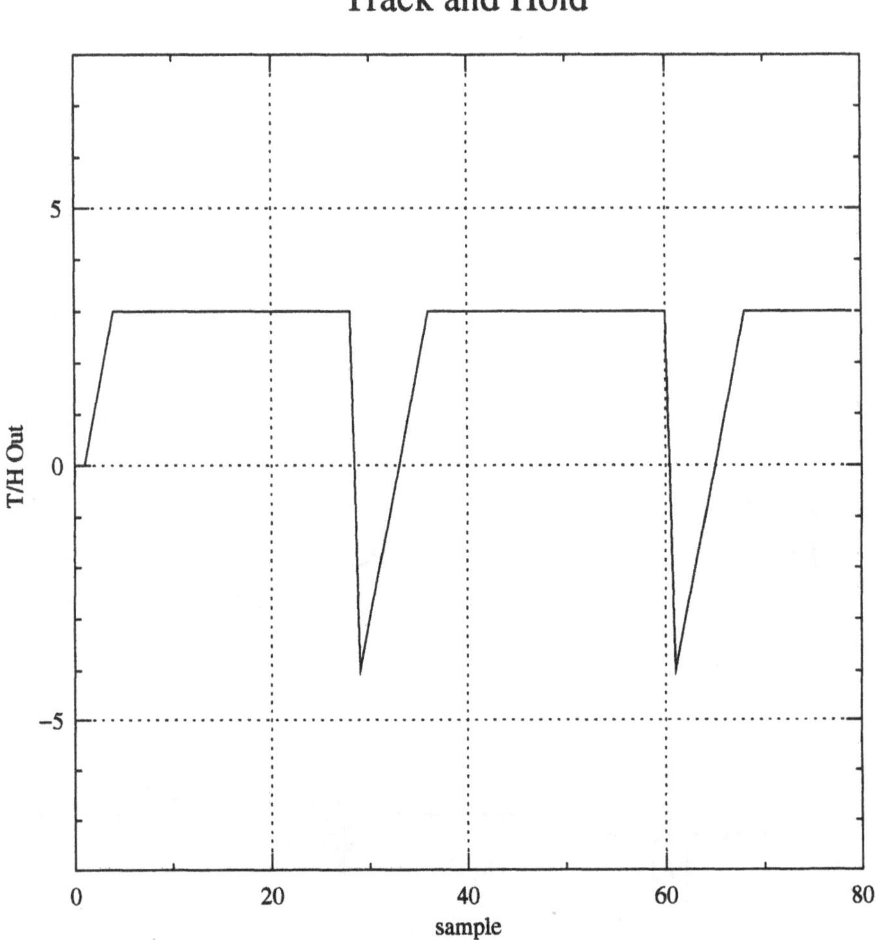

Figure 5.56. Track and Hold Detector Output With a Positive Ramp Input

hold output when the input phase ramp is negative going. The number of bits used by the CPD in general, and the track and hold is particular, is limited to a relatively small number. The reason for this design choice is that the track and hold circuit requires its input signal to have its bits aligned in time. This is required since the MSB's of the input are used to determine the LSB's of the output. This means that the pipeline used by the NCO and phase difference circuit must be determined at the track and hold input. The track and hold output drives a pipelined CIC

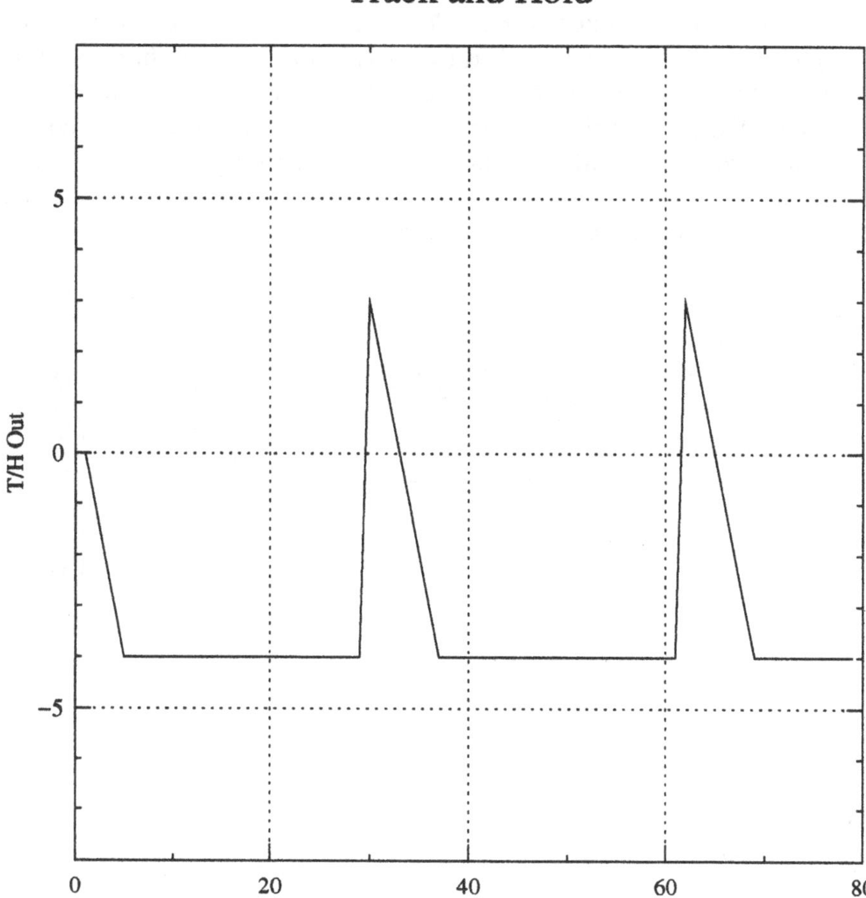

Figure 5.57. Track and Hold Detector Output With a Negative Ramp Input

decimator which means the track and hold output must be skewed prior to entering the pipeline. Since the depth of pipelining is one register per every two bits, we require around $2\left(\frac{N}{2}\right)\left(\frac{N}{2}\right)$ registers[8] for deskewing and skewing the data. If we use the full 10–bits associated with the NCO we will require around 40 flip–flops for this function. However, if we reduce the number of bits to 5 or 6, the number of flip–flops drops to around 10.

[8]The reason the expression is not exact is it doesn't account for the case of N being odd.

5.6.3 Upsampler

The loop filter output is upsampled to match the phase accumulator sample rate. This is accomplished by zero stuffing the loop filter output. A zero stuffing approach is used rather than a zero order hold due to the lower DC gain of the former. The skewing of the bits required for the pipelined adders and phase accumulator is easily combined with the zero stuffing circuit. Figures 5.58 and 5.59 show the schematic for the combined upsampling and skewing circuit.

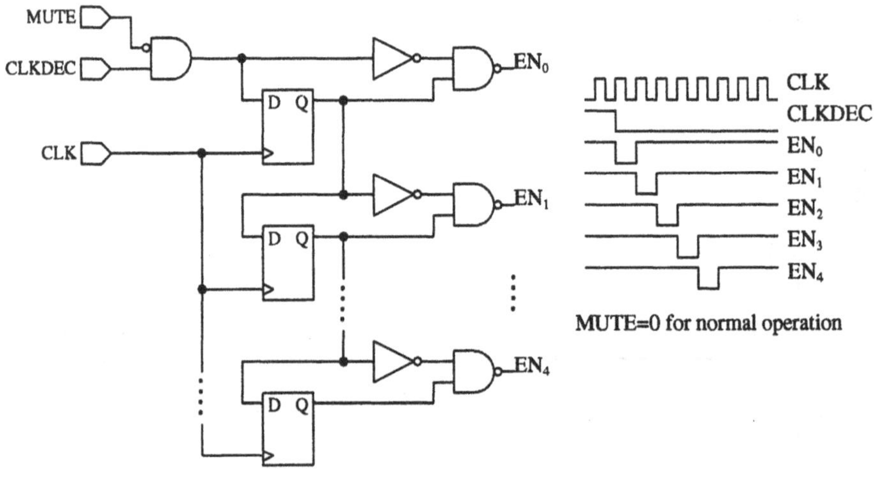

Figure 5.58. Loop Filter Output Skew–Enable Pulse Generation

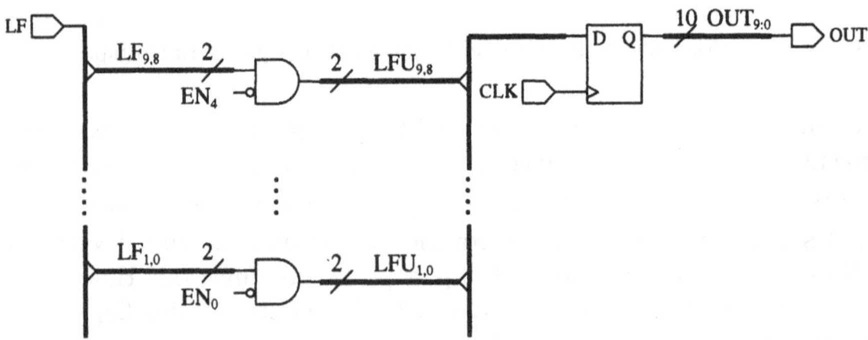

Figure 5.59. Loop Filter Output Skew–Gating Circuit

The circuit operates by gating the loop filter output with a series of enabling pulses. The pulses, shown in Figure 5.60, allow one nonzero sample to be read for each cycle of the low rate clock. The *MUTE* signal in Figure 5.59 allows the loop filter output to be forced to be zero. This feature is used to enable testing of the DPLL open loop response.

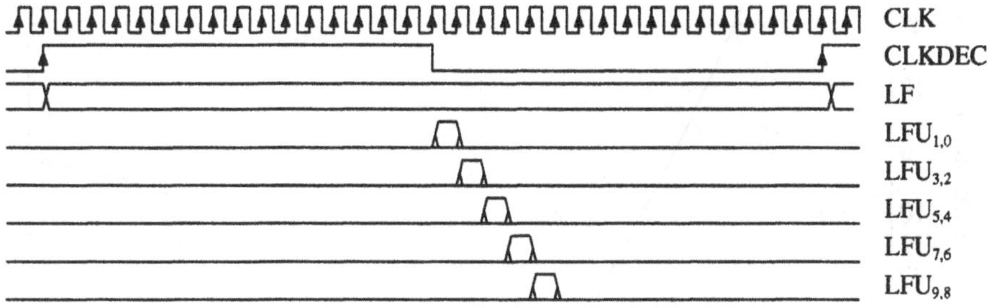

Figure 5.60. Loop Filter Output Skew–Timing Diagram

5.7. Gain Control Multiplier and Error Amp

The final step in the gain error detector is the multiplication of the reference error waveform and the measured phase error. It turns out that the multiplication can be performed with 1 bit of precision. Figures 5.61 and 5.62 show the simulated response of the gain error detector when 1–bit multiplication is used. Note that the scale factor is chosen to produce ± 1 unit levels at the multiplier output. The detector full scale output is somewhat less than the maximum range, but is still large enough. For the purposes of the test chip, the gain control error amplifier was implemented off-chip. This was done to ease the testing of the gain error detector. Figure 5.63 shows the schematic of the off–chip gain control error amplifier. Resistors R2 and R3 produce a mid–supply reference for the integrator. Diode CR1 and resistor R4 prevent the application of negative voltages to the input of the integrated circuit. The error amplifier output voltage is converted to a current by R5 which is connected to the input of an on–chip current mirror. The current input range is $0 \rightarrow 35\mu A$. Diode CR2 prevents the output of the opamp from swinging below its normal minimum output during acquisition[9]. The acquisition time is greatly reduced by not having to wait for the integrator output to swing from the negative rail up to somewhere in its normal operating range.

[9] A condition known as integrator windup [51].

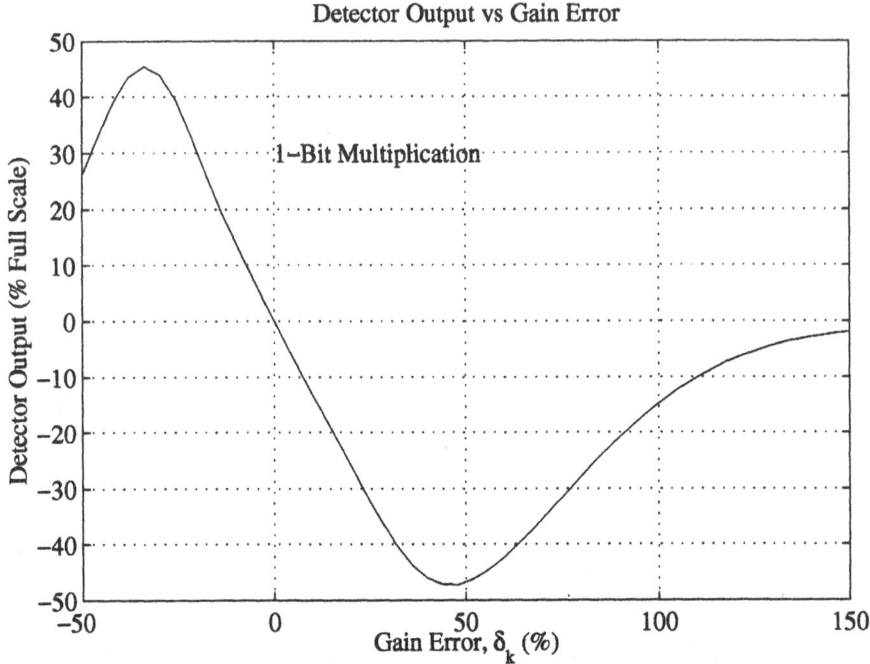

Figure 5.61. Gain Error Detector With One Bit Multiplication – Large Errors

5.8. 1.8 GHz VCO

Figure 5.64 shows the schematic of the integrated circuit voltage controlled oscillator. The inductors are bondwire inductors featuring a quality factor around 12 at 1.8 GHz. Further details on the inductors were given in section 5.2.2. The varactors were discussed in section 5.2.1. The VCO design is heavily based on the bipolar VCO presented in [35]. The layout had to be redone as the bipolar transistor structure had a different layout. In addition, the poly–n+ capacitors used in [35] were replaced with poly–poly capacitors.

5.9. Main Synthesizer

The main modulated Σ–Δ synthesizer is heavily based on the CMOS integrated circuit presented by [24]. That design, which included a 900 MHz input multimodulus divider, phase/frequency detector, loop filter and Σ–Δ modulator, was incorporated into the BiCMOS integrated circuit design and layout. Only minor changes to account for technology differences were made.

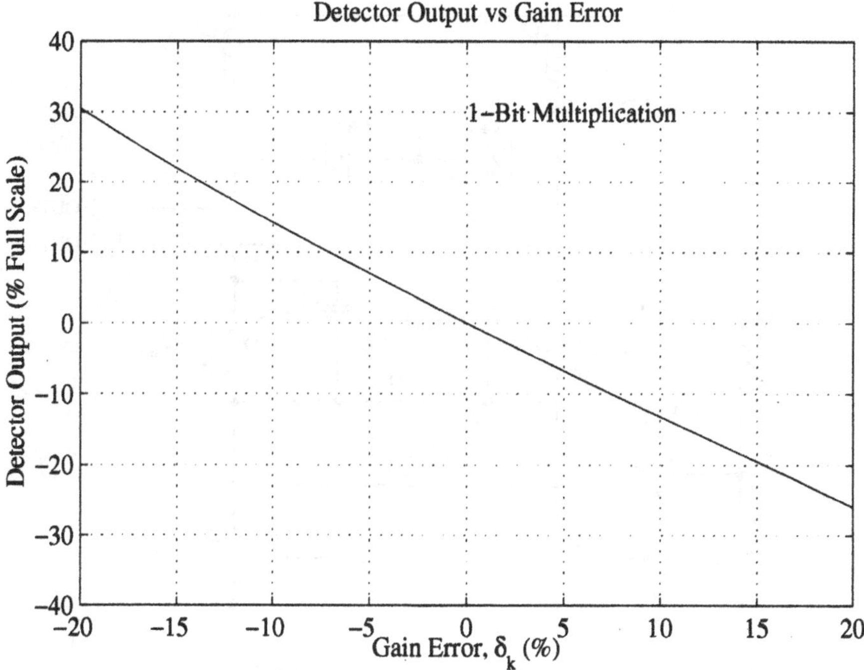

Figure 5.62. Gain Error Detector With One Bit Multiplication - Medium Errors

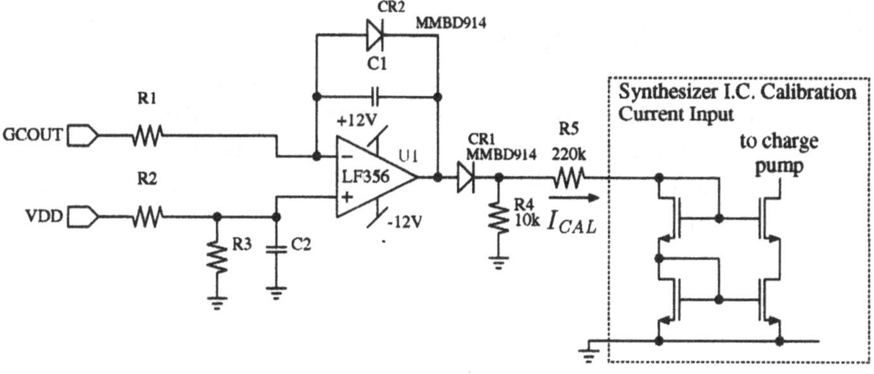

Figure 5.63. Off Chip Gain Control Error Amplifier

The output of the phase/frequency detector in the main synthesizer drives a charge pump whose output drives the loop filter. The current level used by the charge pump is set by the combination of a coarse

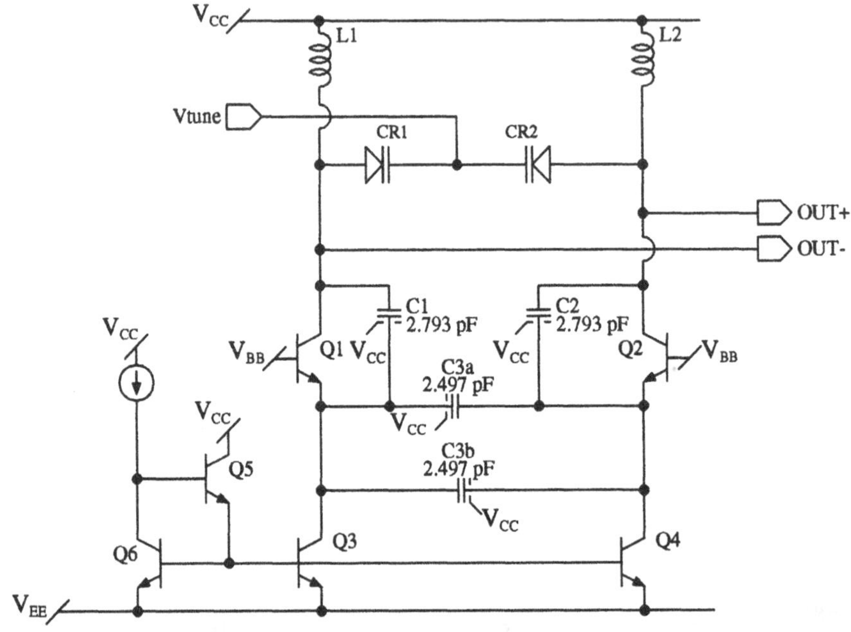

Figure 5.64. Integrated Circuit Voltage Controlled Oscillator

gain control DAC and the analog calibration current which fine tunes the response.

5.10. Σ–Δ Modulator

The Σ–Δ modulator is a second order MASH type structure [52]. Figure 5.65 shows the detailed block diagram of the MASH circuit. The operation of the MASH converter is fairly simple. Consider the first order modulator labeled A. The relationship between the various signals is given by (5.15).

$$x_n + e_{1_{n-1}} = m_{1_n} + e_{1_n} \tag{5.15}$$

Applying the z-transform to (5.15) gives (5.16).

$$M_1(z) = X(z) - (1 - z^{-1})E_1(z) \tag{5.16}$$

From (5.16) we see that the noise transfer function is $(1 - z^{-1})$. The idea behind the MASH converter is to try and cancel or subtract away the first stage quantization error with additional modulators. Turning to the second modulator we see that its output is described by (5.17).

$$M_2(z) = E_1(z) - (1 - z^{-1})E_2(z) \tag{5.17}$$

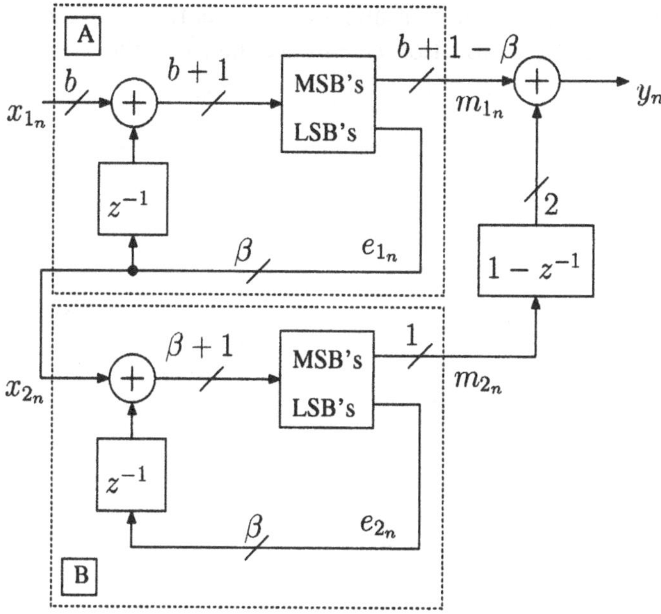

Figure 5.65. Second Order MASH Converter

To use the output of the second modulator to try and cancel the first we simply apply the $(1 - z^{-1})$ transfer function prescribed by (5.16) to M_2 and add it to the output of the first modulator. The results are

$$\begin{aligned} Y(z) &= M_1(z) + (1 - z^{-1})M_2(z) \\ &= X(z) - (1 - z^{-1})E_1(z) + (1 - z^{-1})(E_1(z) - (1 - z^{-1})E_2(z)) \\ &= X(z) - (1 - z^{-1})^2 E_2(z). \end{aligned} \quad (5.18)$$

From (5.18) we see that the quantization noise experiences a second order noise transfer function. If desired, further noise shaping may be achieved by including additional stages. For the application at hand, however, a second order modulator is sufficient.

5.11. Chapter 5 Summary

This chapter has presented two implementations of the new automatic calibration system. The first implementation is a board level implementation using a field programmable gate array to implement all digital functions and discrete transistors and opamps for the analog functions. The second implementation is a full custom BiCMOS integrated circuit. This implementation was discussed in detail. The integrated circuit features 1.8 GHz output center frequency and has the VCO, Σ–Δ synthesizer and automatic calibration circuit on the same die.

Chapter 6

EXPERIMENTAL RESULTS

This chapter presents the measured performance of the automatically calibrated synthesizer. The low frequency discrete prototype was tested using 2 level GFSK at a data rate of 78.125 kbps. Correct operation of the high frequency integrated circuit version was verified using both 2 level and 4 level GFSK modulation. When using 4 level GFSK, a data rate of 5 Mbps is achieved. The inclusion of the coarse digital calibration algorithm discussed in section 4.7 allows the calibration circuit to converge from a wide initial error range. The tested range corresponds to approximately a two to one variation in PLL forward path gain.

6.1. Discrete Prototype Results

This section presents the main results from the discrete prototype of the automatically calibrated Σ–Δ synthesizer.

6.1.1 Discrete Prototype GFSK/GMSK Eye Pattern and Spectrum

To test modulator performance, a Hewlett Packard 89440A Vector Signal Analyzer was used to demodulate the RF output of the test system. Figure 6.1 shows the demodulated eye pattern with the automatic calibration circuit disabled. The coarse calibration DAC is set to produce a forward path gain in the PLL which is too low. The resulting intersymbol interference (ISI) can be seen. Figure 6.2 shows the RF output spectrum that results from the eye pattern in Figure 6.1. The narrow spectrum is indicative of the modulation index being too low. When the coarse gain control DAC is set to produce a forward path gain in the PLL which is too high, the eye pattern in Figure 6.3 and

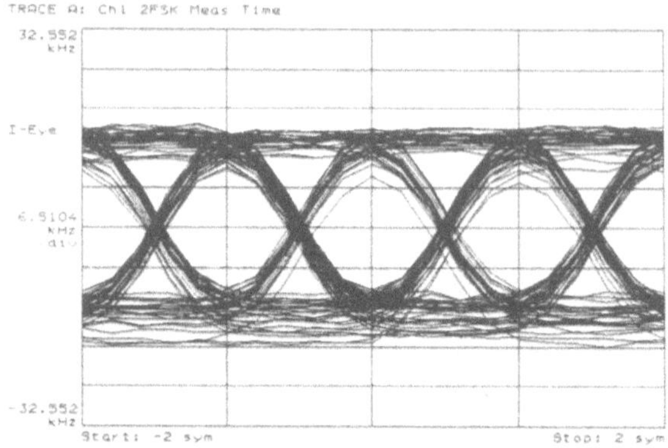

Figure 6.1. Demodulated 78.125 kbps Eye Pattern – Gain Too Low

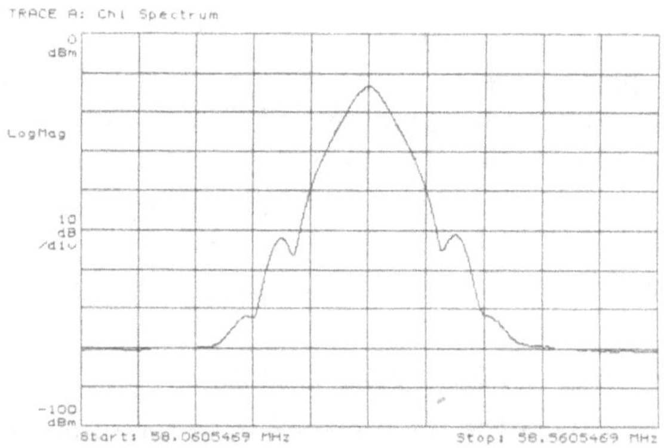

Figure 6.2. Measured 78.125 kbps RF Spectrum – Gain Too Low

corresponding spectrum in 6.4 result. As in the low gain case, ISI is visible in the eye pattern. In the high gain case, the spectrum is wider and has a flatter top which is indicative of the modulation index being too high.

Figure 6.5 shows the measured eye pattern when the automatic calibration circuit is enabled. With automatic calibration enabled, the eye pattern remains unchanged when the coarse gain control DAC is varied. Figure 6.6 shows the measured automatically calibrated RF output spectrum.

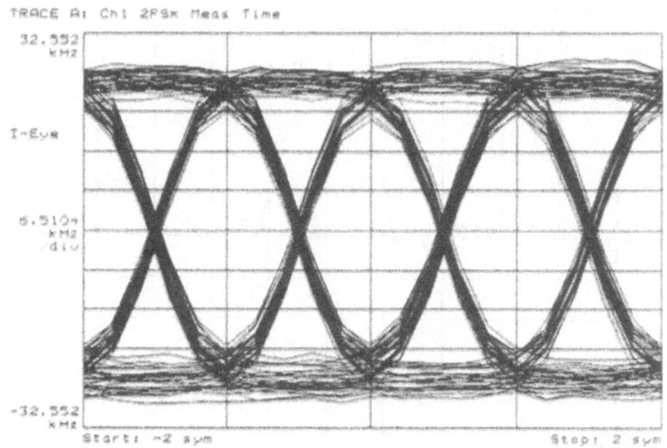

Figure 6.3. Demodulated 78.128 kbps Eye Pattern – Gain Too High

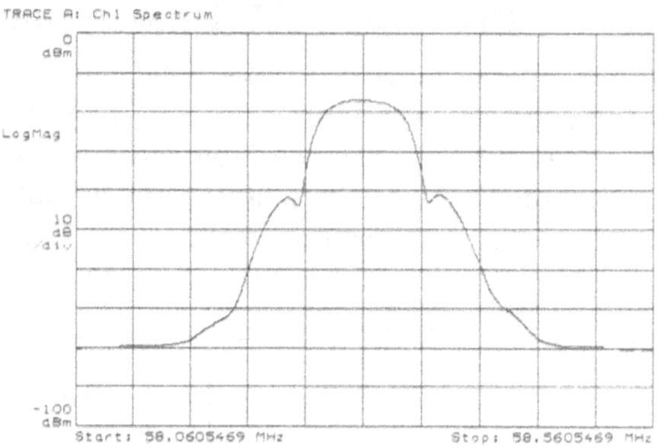

Figure 6.4. Measured 78.125 kbps RF Spectrum – Gain Too High

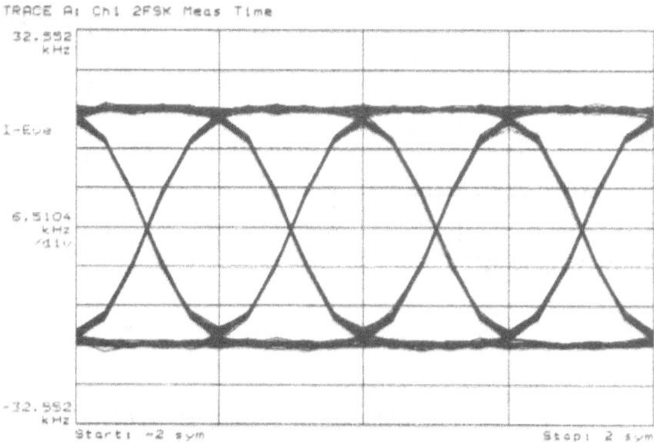

Figure 6.5. Demodulated 78.125 kbps Eye Pattern – Autocal Enabled

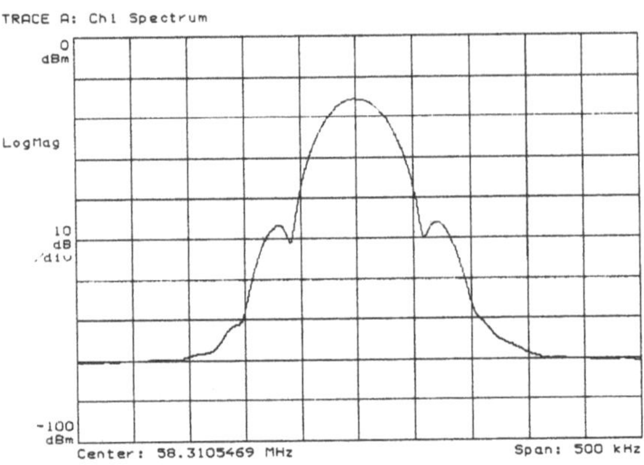

Figure 6.6. Measured 78.125 kbps RF Spectrum – Autocal Enabled

6.1.2 Discrete Prototype RF Phase Quantizer

To verify correct operation of the phase quantizer the modulation was turned off in the main synthesizer. The 2-bit samples from the phase quantizer were sent to a discrete 2-bit DAC and observed with an oscilloscope. Figure 6.7 shows the phase quantizer output when the fractional frequency of the main synthesizer is set to -10.9875 kHz. The

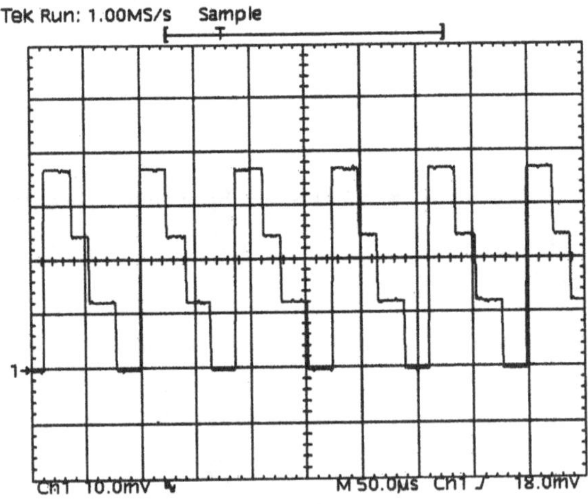

Figure 6.7. Measured Discrete Prototype Phase Quantizer Output

frequency of the RF phase quantizer output is around 11 kHz as anticipated. Figure 6.8 shows the same signal with the fractional frequency set to +10.9875 kHz. As expected, the frequency is the same, but the direction of the phase ramp has changed sign.

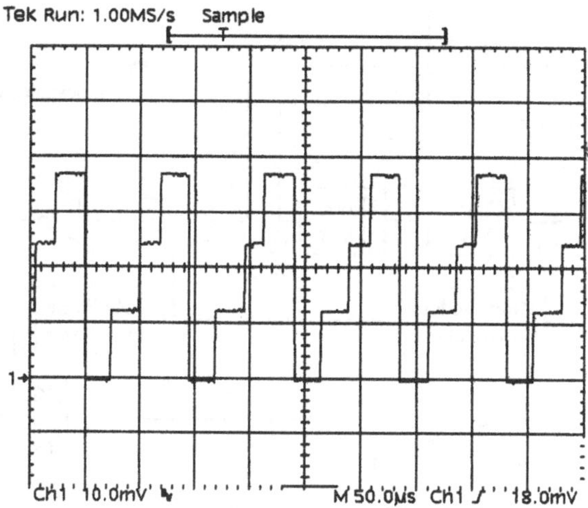

Figure 6.8. Measured Discrete Prototype Phase Quantizer Output

6.2. Integrated Circuit Test Environment

This section describes some of the features of the integrated circuit and the test board which were included to enhance the testability of the chip. The integrated circuit provides access to several of the key signals in the calibration system. Figure 6.9 shows a block diagram of the chip which includes these signals. Not shown in Figure 6.9 is a serial input configuration register that is used to program the center frequency and control the various test multiplexers. The test mux, *U2*, can be programmed to bring one of five key digital signals off the chip. In addition, *U1*, can be used to select an off chip source for the RF phase quantizer samples. The flexibility in providing stimulus and observing internal signals allows the digital phase locked loop to be tested independently from the rest of the chip. Under control of the serial configuration registers the DPLL loop filter output can be forced to zero to be able to examine the DPLL open loop response. The other digital signals which are brought off chip for test purposes are the RF phase quantizer output, the two one bit inputs to the gain error detector multiplier, the multiplier output and the multimodulus divider output. In terms of analog signals, the main synthesizer loop filter is connected to the VCO externally. The external connection allows direct observation of the transmit eye pattern. Also the VCO input may be manually swept to measure the VCO tuning curve.

Figure 6.10 shows a simplified block diagram for the test board. The test board includes a variety of options for providing inputs to the chip as well as processing some of the test outputs. All of the digital inputs

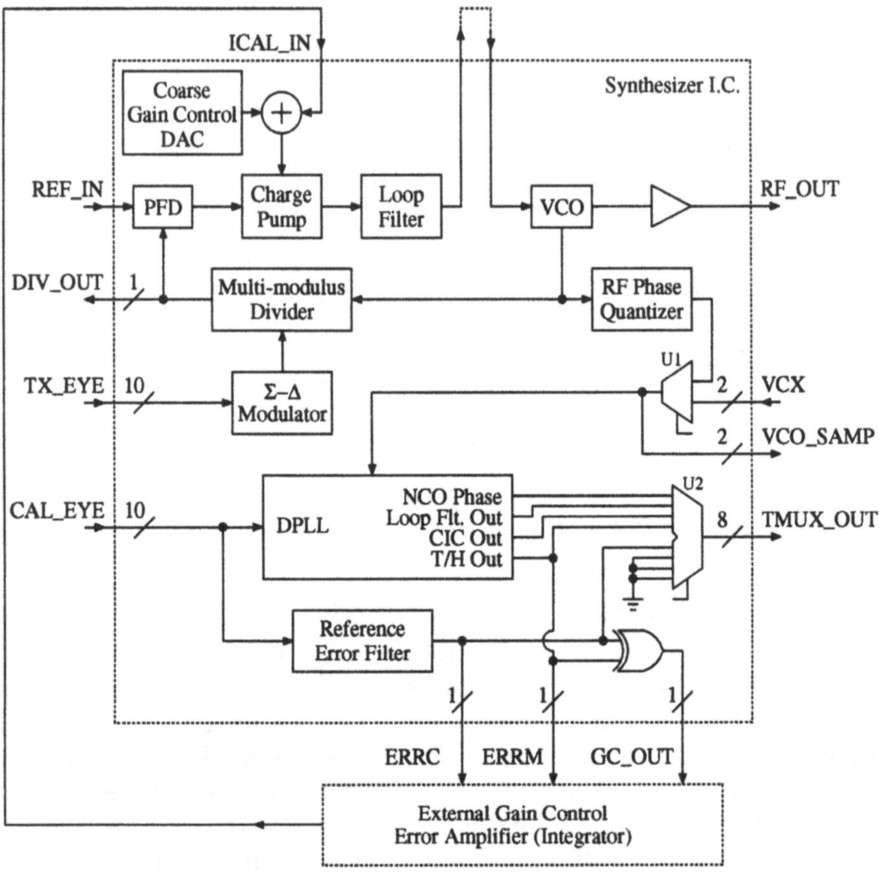

Figure 6.9. Integrated Circuit Block Diagram – Testability Features

and outputs on the synthesizer chip are routed to a field programmable gate array (FPGA). The FPGA can be programmed to configure the system in many different ways. The various digital inputs to the synthesizer chip may be generated directly by the FPGA, read from the A/D converter (allowing the use of a signal generator as an input source), or read from RAM. An 8–bit D/A converter is used to convert the synthesizer test multiplexer output to analog form so waveforms may be easily displayed on an oscilloscope. A pair of 2–bit D/A converters are available to display the sampled and quantized RF output phase along with a reference version generated by the FPGA. Several line drivers and receivers are available to provide or accept various clock, data and trigger signals.

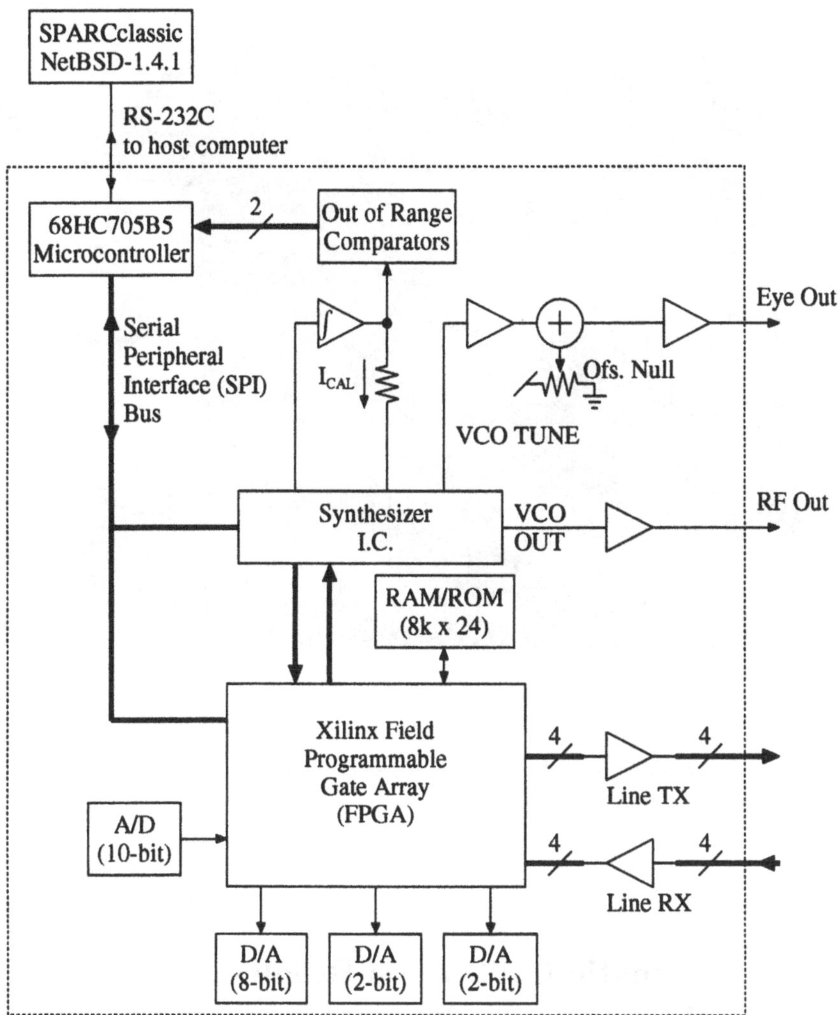

Figure 6.10. Test Board Block Diagram

The operation of the test board is controlled by a computer which communicates with the board via a RS–232 serial interface. An on-board microcontroller implements the asynchronous serial interface and provides a simple command line interface for accessing the serial peripheral interface (SPI) which communicates with the FPGA and synthesizer chip. A program on the host computer provides the user interface and generates the lower level commands which are sent to the test board. Figure 6.11 contains a photograph of the synthesizer chip test board. The modulated synthesizer integrated circuit is marked in the photo as

Figure 6.11. Photo of Synthesizer Chip Test Board

U1.

6.3. Automatic Calibration Results

6.3.1 GFSK/GMSK Eye Pattern and Spectrum

To test modulator performance, a Hewlett Packard 89440A Vector Signal Analyzer was used to demodulate the RF output of the test system. Figure 6.12 shows the demodulated eye pattern with the automatic calibration circuit disabled. The coarse calibration DAC is set to produce a forward path gain in the PLL which is too low. The resulting intersymbol interference (ISI) can be seen. Figure 6.13 shows the RF output spectrum that results from the eye pattern in Figure 6.12. The narrow spectrum is indicative of the modulation index being too low. When the coarse gain control DAC is set to produce a forward path gain in the PLL which is too high, the eye pattern in Figure 6.14 and corresponding spectrum in 6.15 result. As in the low gain case, ISI is visible in the eye pattern. In the high gain case, the spectrum is wider

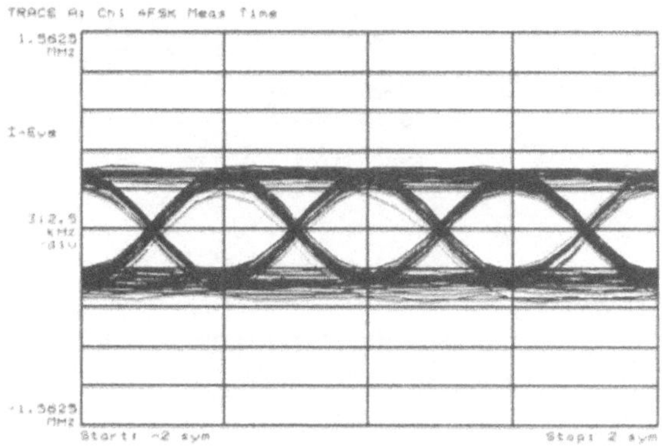

Figure 6.12. Demodulated 2.5 Mbps Eye Pattern – Gain Too Low

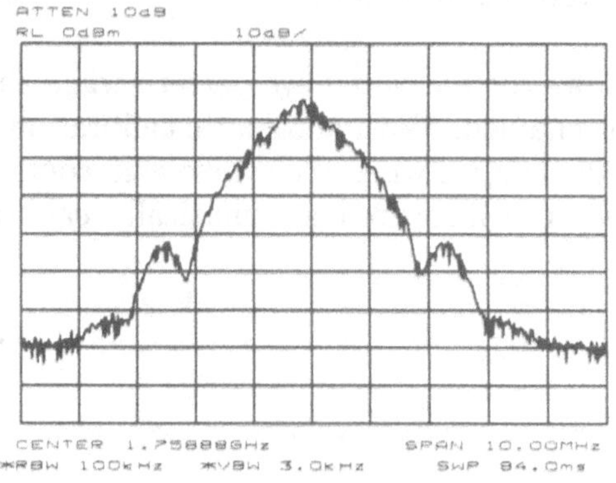

Figure 6.13. Measured 2.5 Mbps RF Spectrum – Gain Too Low

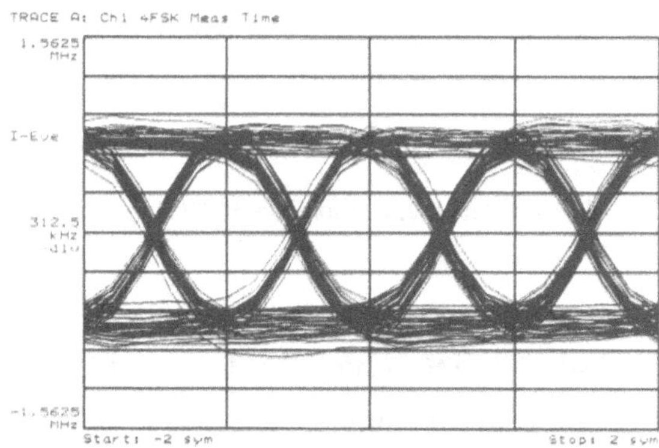

Figure 6.14. Demodulated 2.5 Mbps Eye Pattern – Gain Too High

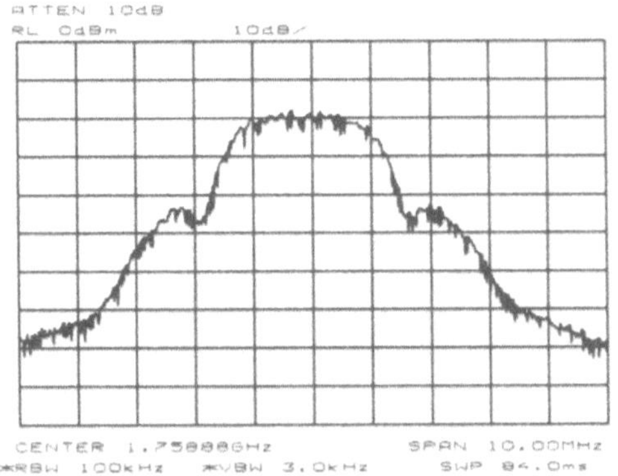

Figure 6.15. Measured 2.5 Mbps RF Spectrum – Gain Too High

and has a flatter top which is indicative of the modulation index being too high.

Figure 6.16 shows the measured eye pattern when the automatic calibration circuit is enabled. With automatic calibration enabled, the eye pattern remains unchanged when the coarse gain control DAC is varied. Figure 6.17 shows the measured automatically calibrated RF output

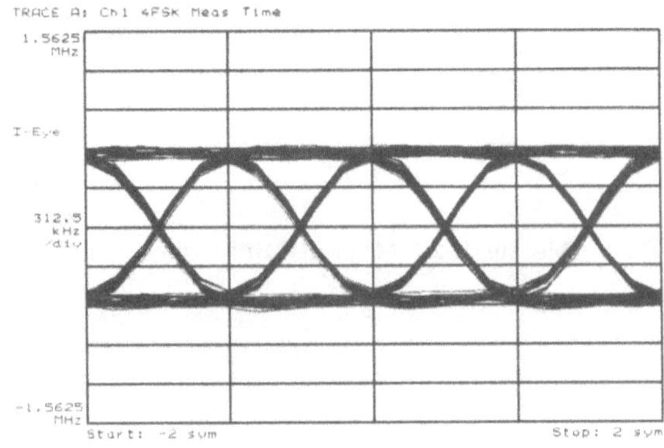

Figure 6.16. Demodulated 2.5 Mbps Eye Pattern – Autocal Enabled

spectrum.

6.3.2 4 Level GFSK Eye Pattern and Spectrum

The presented calibration technique is not tied to 2 level modulation types. To verify correct operation with a higher order modulation type, the transmit waveform was programmed to be a 4 level GFSK type. Figure 6.18 shows the demodulated 4-GFSK eye pattern when the gain

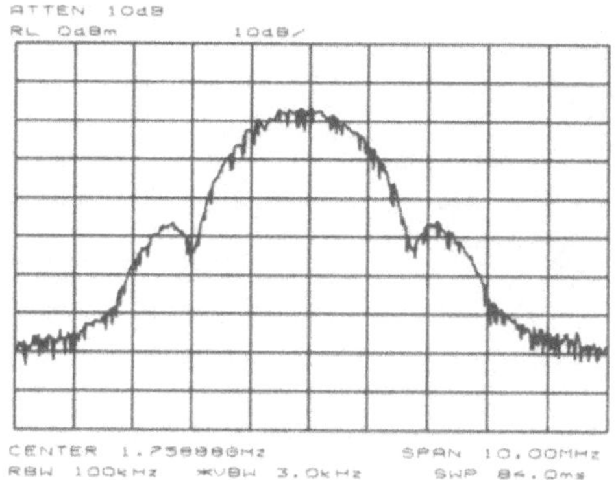

Figure 6.17. Measured 2.5 Mbps RF Spectrum – Autocal Enabled

is too low. A large amount of ISI is present. The demodulated eye

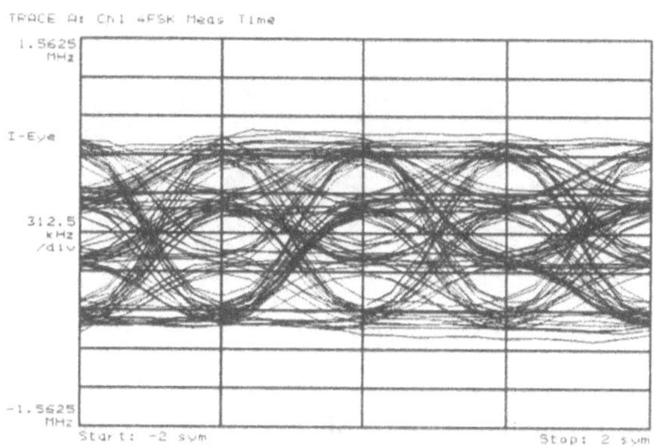

Figure 6.18. Demodulated 4-GFSK Eye Pattern – Gain Too Low

pattern in Figure 6.19 results when the gain is too high. Again a large amount of ISI is present. When the automatic calibration circuit is enabled, the eye pattern in Figure 6.20 is obtained regardless of the initial gain setting.

6.3.3 Calibration Dynamics

The disturbances which the automatic calibration circuit must track are thermal variations. Correspondingly, the loop bandwidth of the calibration circuit is set to be just over 1 Hz. Figure 6.21 and Figure 6.22 show the response of the automatic calibration circuit to a step change in the setting of the coarse calibration DAC.

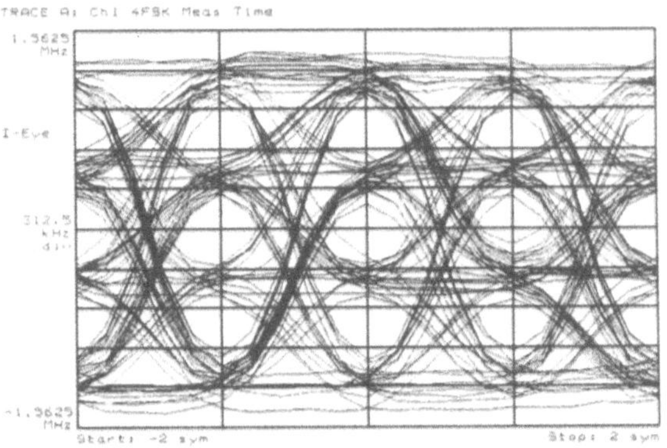

Figure 6.19. Demodulated 4-GFSK Eye Pattern – Gain Too High

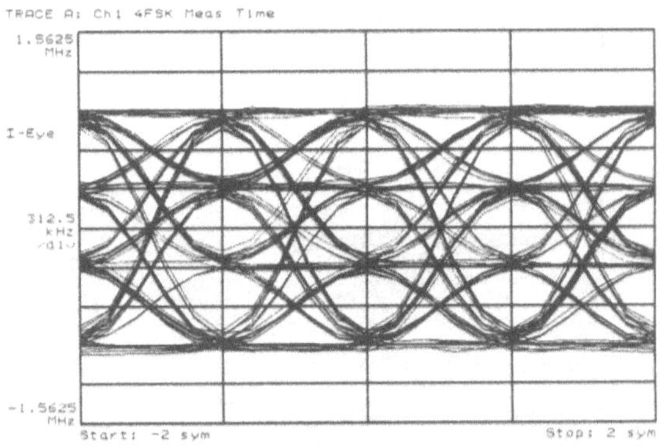

Figure 6.20. Demodulated 4-GFSK Eye Pattern – Autocal Enabled

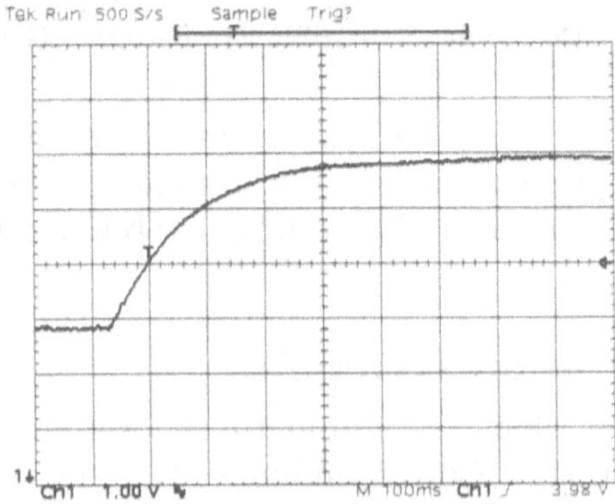

Figure 6.21. Automatic Calibration Step Response – Positive Step (100 ms/div)

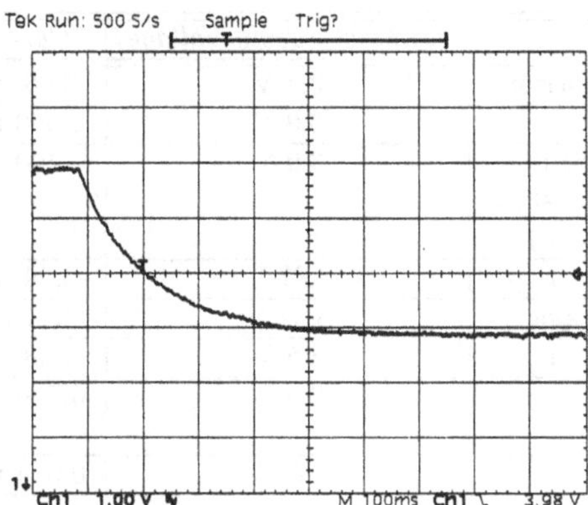

Figure 6.22. Automatic Calibration Step Response – Negative Step (100 ms/div)

6.4. Power Consumption

Table 6.1 lists the measured power consumption for the test chip. Although the DPLL, reference error filter, and Σ–Δ modulator were designed to operate at 1.5 Volts, they were measured at 3.0 Volts due to a bug in some level conversion circuitry. However, these sections of the chip represent a small portion of the total power. Operation of these 3 blocks at 1.5 Volts was verified, but not used for testing of the complete chip due to the aforementioned level conversion mistake. Separate supply pins were used for the individual circuit blocks to allow individual power consumption measurements. In practice, a smaller number of supply voltages would be used. Due to a biasing error the VCO and prescaler were not run at the desired 3.0 Volts supply. Figure 6.23 shows the relative power consumption of the various circuits in the synthesizer chip.

6.5. VCO performance

This section presents the measured performance of the voltage controlled oscillator. Figure 6.24 shows the measured VCO output frequency as a function of the control voltage. Figure 6.25 shows the associated VCO gain. The VCO gain was measured at several DC operating points in a closed loop configuration. To make the measurement, a high resolution voltmeter was used to measure the tuning voltage. The synthesizer output was then reprogrammed using a small step size and the resulting change in tuning voltage measured.

The phase noise of the VCO was measured under several conditions. These measurements are intended to characterize both the VCO performance and any degradation due to coupling of noise from the digital

Circuit	*Nominal Supply Voltage*	*Current*	*Power*
RF Phase Quantizer	3.3 V	3.98 mA	13.2 mW
DPLL	3.0 V	0.558 mA	1.67 mW
Reference Error Filter (includes pipeline skew shared by DPLL)	3.0 V	0.490 mA	1.46 mW
VCO	2.87 V	3.54 mA	10.2 mW
Bipolar Prescaler	3.22 V	4.11 mA	13.2 mW
CMOS Divider	3.5 V	9.48 mA	32.3 mW
Σ-Δ Modulator	3.0 V	0.586 mA	1.75 mW
Loop Filter	3.29 V	0.674 mA	2.22 mW
Misc.			2.13 mW
Total			78.0 mW

Table 6.1. Measured Power Consumption

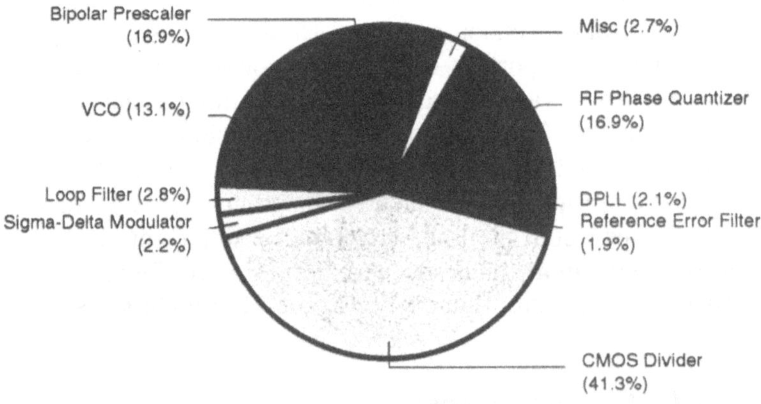

Figure 6.23. Breakdown of Power Consumption

circuitry. In all cases, the VCO tuning voltage was set with a low noise source. Figure 6.26 shows the measured VCO phase noise with and without all of the digital circuits disabled. The difference in the measured phase noise level between the two cases is within the measurement accuracy of the instrument. From this plot we conclude that the phase noise within 10 MHz of the carrier is unaffected by the presence of the

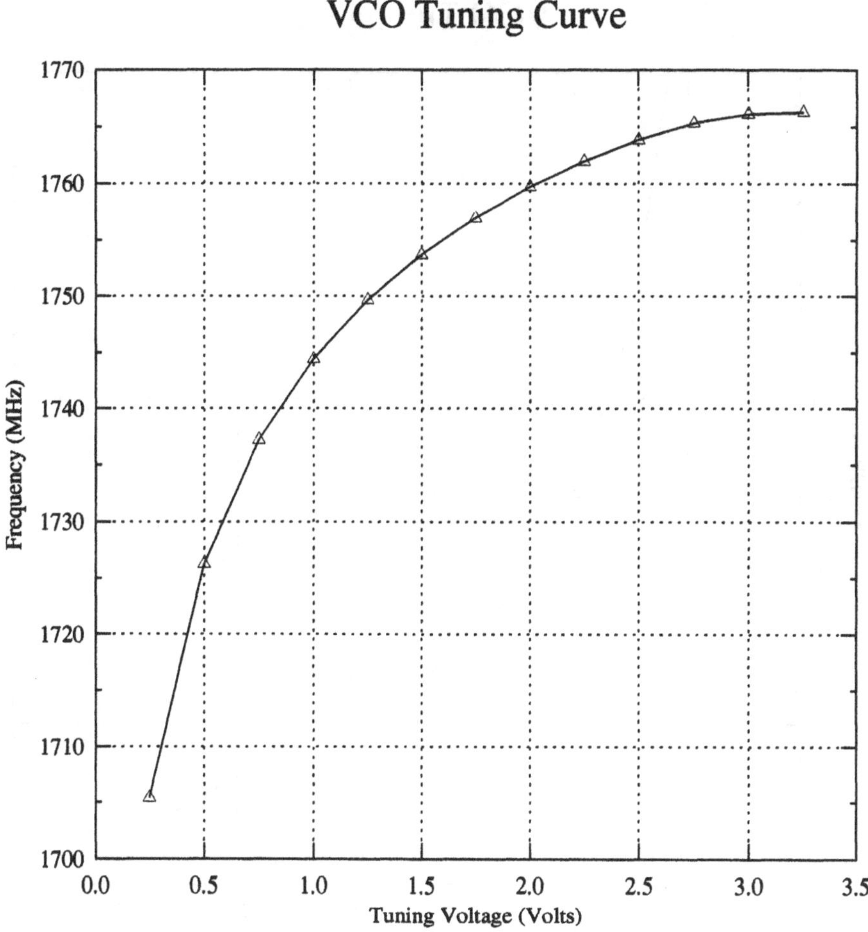

Figure 6.24. Measured VCO Tuning Curve

Offset Frequency	*Phase Noise*
10 kHz	-72 dBc/Hz
100 kHz	-92 dBc/Hz
1 MHz	-112 dBc/Hz

Table 6.2. Measured VCO Phase Noise

digital circuits on the chip and on the test board. Table 6.2 summarizes the phase noise performance of the VCO.

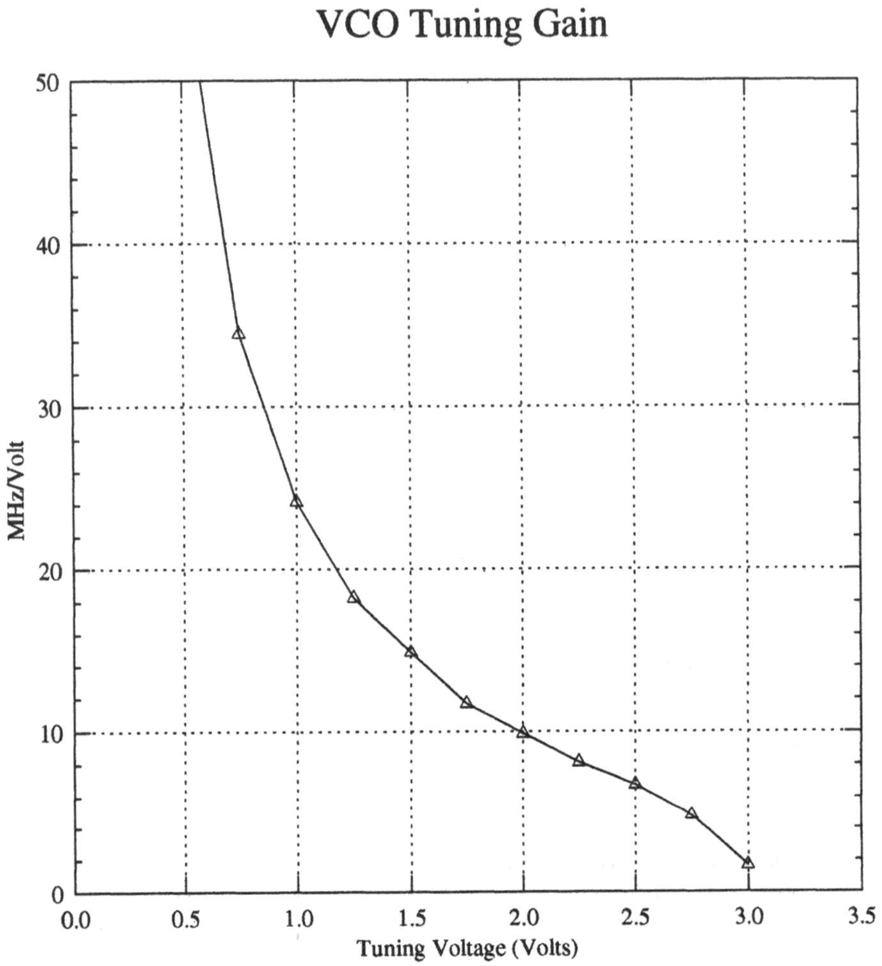

Figure 6.25. Measured VCO gain

Figure 6.27 shows the measured RF output spectrum with the digital portion of the chip disabled. Figure 6.28 shows the output spectrum with a wider span under the same conditions as in Figure 6.27.

When the digital circuitry is enabled, a few spurious outputs are observed. To assist in isolating the origin of the spurious outputs, the connection from loop filter output to VCO control input was broken. A low noise voltage source was used to tune the VCO. Figure 6.29 shows the measured RF output spectrum under these conditions. Figure 6.30 shows the measured output spectrum with a wider span under the same

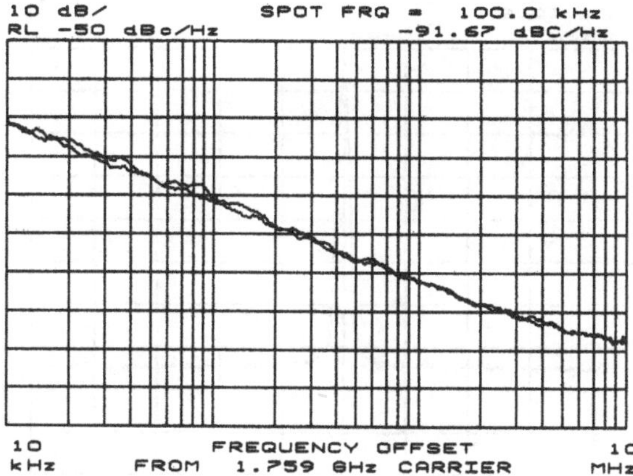

Figure 6.26. Measured Phase Noise - With and Without Digital Circuits Enabled

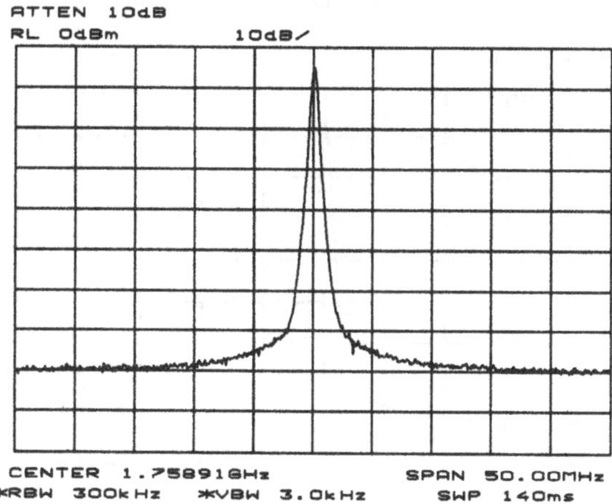

Figure 6.27. Measured VCO Output Spectrum–Digital Circuits Disabled

test conditions as in Figure 6.29. The source of the spurious output at $\pm n * 20$ MHz offsets from the carrier was determined to be the RF phase quantizer. This determination was made by observing that with all of the digital circuits enabled, the spurious outputs could be eliminated by turning off the bias current to the RF latches. The latch in the RF phase quantizer (refer to the schematic in Figure 5.26) presents a different input impedance in its transparent state than in its latched state. Since the circuitry in between the oscillator resonator and the latch input, a cascade of two emitter followers and the quadrature network, does not have perfect reverse isolation, the resonator impedance is disturbed by the latch turning on and off. To estimate the amount of frequency

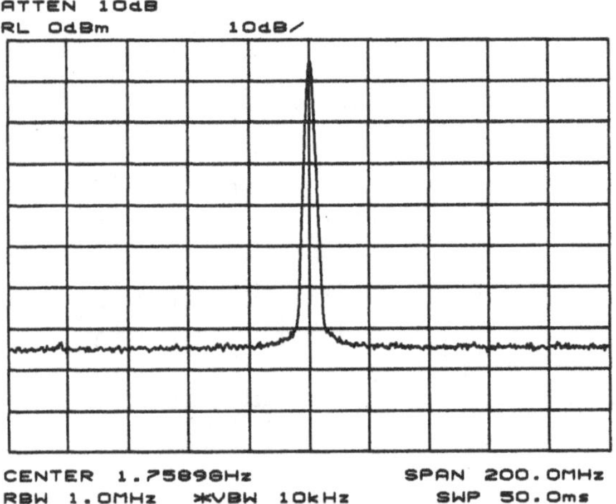

Figure 6.28. Measured VCO Output Spectrum–Digital Circuits Disabled

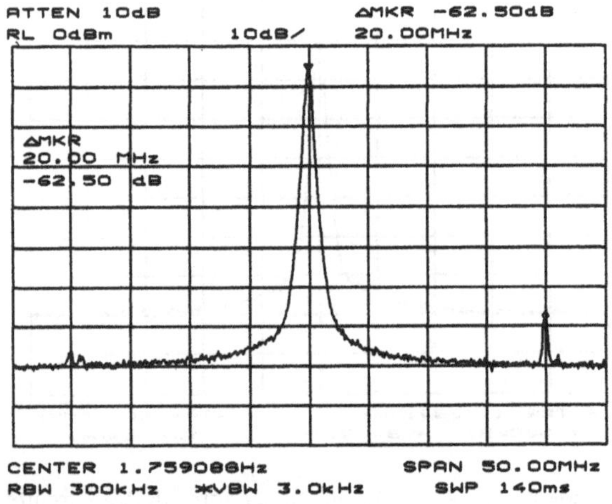

Figure 6.29. Measured VCO Output Spectrum–Digital Circuits Enabled

change that results, consider the natural frequency of a L–C resonator when the capacitance is disturbed by an amount Δ_C. The undisturbed natural frequency is

$$f_0 = \frac{1}{2\pi\sqrt{LC}}. \tag{6.1}$$

When the capacitance in changed by Δ_C the new frequency is

$$f_1 = \frac{1}{2\pi\sqrt{L(C+\Delta_C)}} = \frac{1}{2\pi\sqrt{LC}\sqrt{1+\frac{\Delta_C}{C}}} = f_0\frac{1}{\sqrt{1+\frac{\Delta_C}{C}}}. \tag{6.2}$$

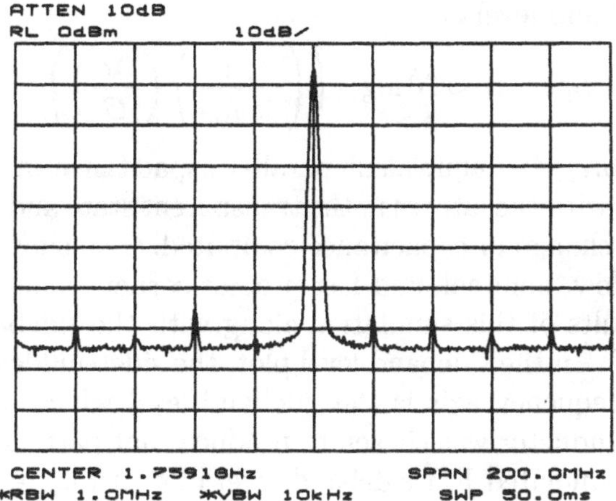

Figure 6.30. Measured VCO Output Spectrum–Digital Circuits Enabled

The total frequency shift is

$$\Delta_f = f_1 - f_0 = \frac{1}{2\pi\sqrt{LC}}\left(\frac{1}{\sqrt{1+\frac{\Delta_C}{C}}} - 1\right) = f_0\left(\frac{1}{\sqrt{1+\frac{\Delta_C}{C}}} - 1\right). \tag{6.3}$$

Assuming that the fractional change in capacitance is very small, $\frac{\Delta_C}{C} \ll 1$, we can expand (6.3) in a first order Taylor series to obtain

$$\Delta_f \approx -\frac{1}{2} f_0 \frac{\Delta_C}{C}. \tag{6.4}$$

Since the spurious frequency modulation of the VCO is small, we can calculate the sideband level under the narrowband FM approximation. Defining the phase modulation index [53], m_θ as the ratio of peak frequency deviation to modulating frequency and recalling that the fundamental frequency component of a square wave is $\frac{4}{\pi}$ times the square wave amplitude, we obtain

$$m_\theta = \frac{\frac{4}{\pi}\frac{|\Delta_f|}{2}}{f_{ref}} = \frac{2}{\pi}\frac{|\Delta_f|}{f_{ref}}. \tag{6.5}$$

The sideband level relative to the carrier will be $20\log_{10}\left(\frac{m_\theta}{2}\right)$. Substitution of (6.4) into (6.5) gives

$$m_\theta = \frac{1}{\pi}\frac{f_0}{f_{ref}}\frac{\Delta_C}{C}. \tag{6.6}$$

The final sideband level is

$$Spur_{FM} \approx 20\log_{10}\left(\left(\frac{f_0}{2\pi f_{ref}}\right)\left(\frac{\Delta_C}{C}\right)\right). \tag{6.7}$$

To evaluate (6.7), the equivalent parallel capacitance in the resonator was simulated with the latches in the transparent state and in the latched state and the change in capacitance computed. The simulation was repeated with slow, nominal, and fast process corner models. Figure 6.31 shows the results of this simulation along with the sideband level predicted by (6.7). In the sideband level plot, the sideband level for a given value of the frequency axis is the sideband level which would result if the resonator inductor was chosen to produce that particular center frequency. The simulated FM sideband level plot in Figure 6.31 suggests

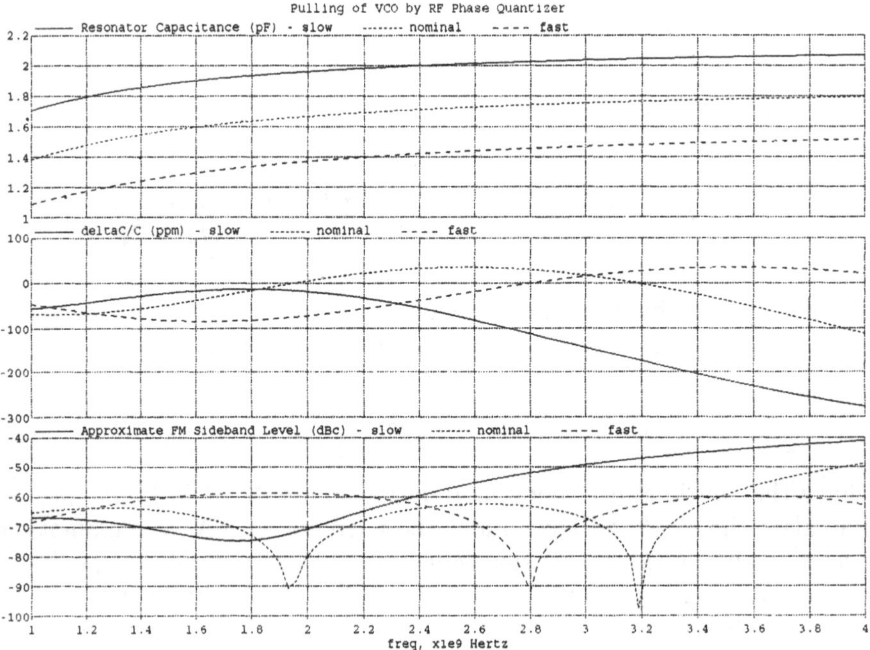

Figure 6.31. Simulated Pulling of VCO by the RF Phase Quantizer

that the observed spurious output level is consistent with pulling of the VCO by the RF phase quantizer. Some possible solutions to this problem are the use of dummy latches which are driven out of phase with the primary latches to maintain constant impedance and the use of buffer stages with better reverse isolation. The latter is likely to be preferred on a power consumption basis.

The astute reader will note that the measured spurious sidebands in Figure 6.29 do not have identical amplitudes. This asymmetry indicates that the sidebands are due to both amplitude and frequency modulation. A transient simulation of the oscillator produced an amplitude variation level corresponding to an AM sideband of approximately -65 to -70 dBc. This is close to the level which would be required to produce the amount of asymmetry which was observed. Due to the long simulation time for transient solutions where the desired tolerance is consistent with resolving -70 dB variations, only a few data points were gathered. In addition, simulations of oscillator amplitudes are typically not all that accurate due to modeling errors. The conclusion is that the impedance variation of the RF latches is probably also responsible for the AM sideband generation as well as the FM sideband generation.

6.6. RF Phase Quantizer

The parameters of interest related to the RF phase quantizer are linearity and noise. The linearity of the phase quantizer is set by a combination of factors. The primary sources of nonlinearity are errors in the quadrature network and input offset of the latches. These two nonidealities both modify the phase where the output code transitions. In section 5.1.5 it was shown that the calibration circuit is relatively insensitive to nonlinearities in the phase quantizer.

To verify correct operation of the phase quantizer and also of the DPLL, the modulation was turned off in the main synthesizer. The on chip test multiplexer was used to bring the numerically controlled oscillator (NCO) phase off chip. In addition, the 2-bit samples from the phase quantizer are available off chip. These two signals were sent to a pair of DAC's and observed with an oscilloscope. Figure 6.32 shows the phase quantizer output (lower trace) and the NCO phase (upper trace) when the fractional frequency of the main synthesizer is set to -1.0839844 MHz. The two signals are indeed phase locked to each other and the frequency is 1.08 MHz as anticipated. Figure 6.33 shows the same waveforms with the fractional frequency set to +1.0839844 MHz. As expected, the frequency is the same, but the direction of the phase ramps has changed sign.

Figure 6.34 shows a noise model for the synthesizer and RF phase quantizer. The primary sources of noise are: noise in the charge pump current source, VCO noise, quantization noise from the Σ–Δ modulator, and quantization noise from the RF phase quantizer. Figure 6.35 shows the relative contributions to the noise at the phase quantizer output from each of the sources shown in Figure 6.34. There are two distinct regions of operation. Inside the synthesizer loop bandwidth, the noise

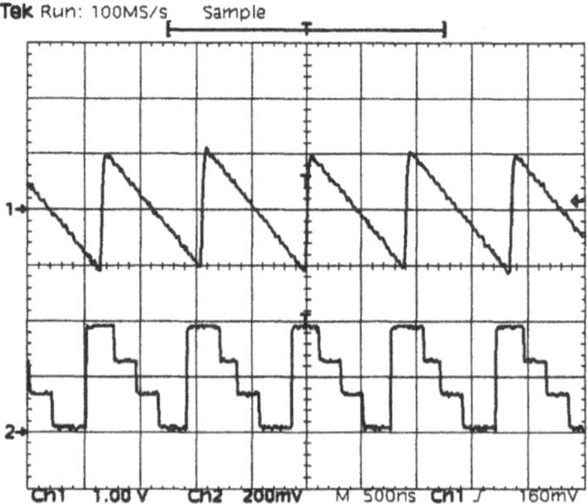

Figure 6.32. Measured Phase Quantizer Output and Phase Locked NCO Output

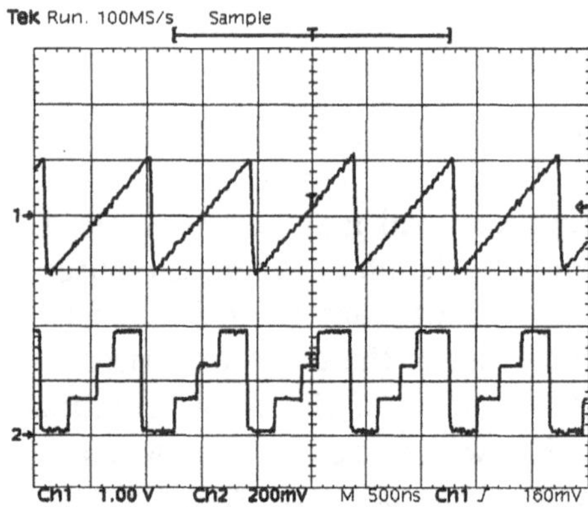

Figure 6.33. Measured Phase Quantizer Output and Phase Locked NCO Output

is dominated by the charge pump current source noise[1]. Outside the synthesizer loop bandwidth, the noise drops until it hits the noise floor imposed by the phase quantizer. Figure 6.36 shows the measured phase quantizer output noise along with the noise predicted by the model in Figure 6.34. Good agreement is observed which indicates that additional noise sources, such as clock jitter in the phase quantizer sample clock, do not limit the noise performance of the phase quantizer.

6.7. Summary

This chapter has presented the main experimental results from the fabricated integrated circuit. The full custom test chip includes the Σ–

[1]This noise is simply the thermal and $\frac{1}{f}$ noise of the devices used in the current source.

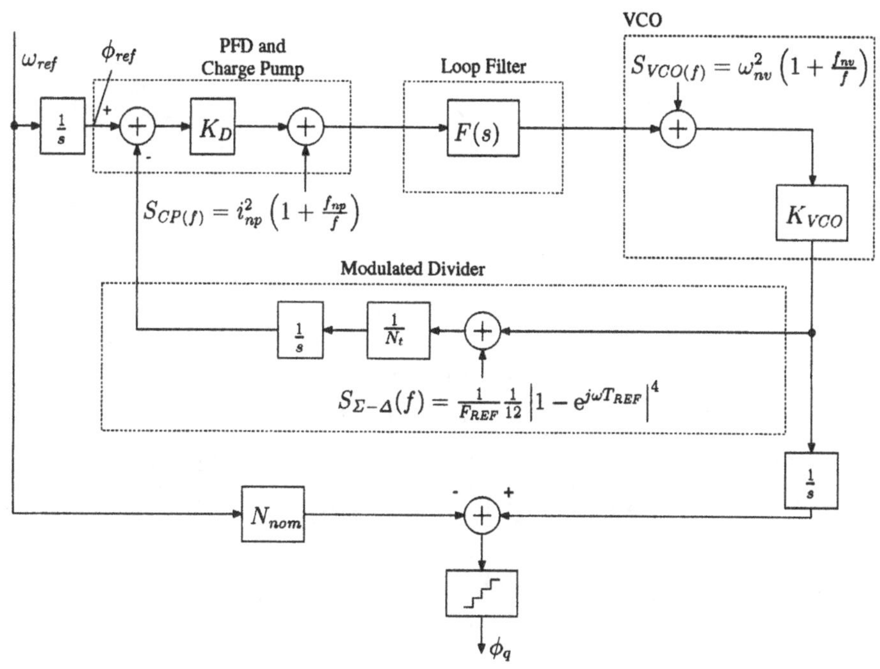

Figure 6.34. System for Calculating RF Phase Quantizer Output Noise

Specification	*Value*
Data Rate	2.5 Mbps using 2 level GFSK 5.0 Mbps using 4 level GFSK
Output Frequency Range	1.705 GHz to 1.765 GHz
Power Consumption	78 mW
Die Area	4.7 mm^2
MOS Transistors	14982
Bipolar Transistors	49
Resistors	171
Capacitors	30
Varactor Diodes	3
Bondwire Inductors	2

Table 6.3. Summary of Test Chip Results

Δ modulator, multimodulus divider, VCO, loop filter and automatic calibration circuit.

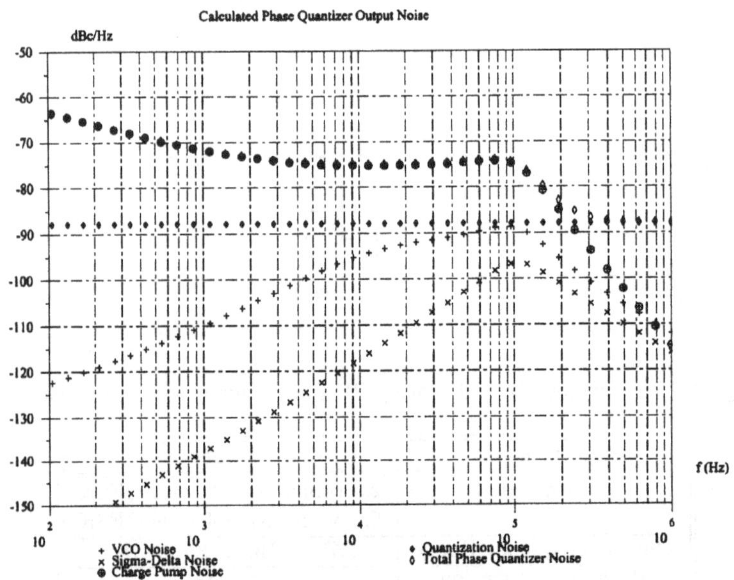

Figure 6.35. Calculated RF Phase Quantizer Output Noise

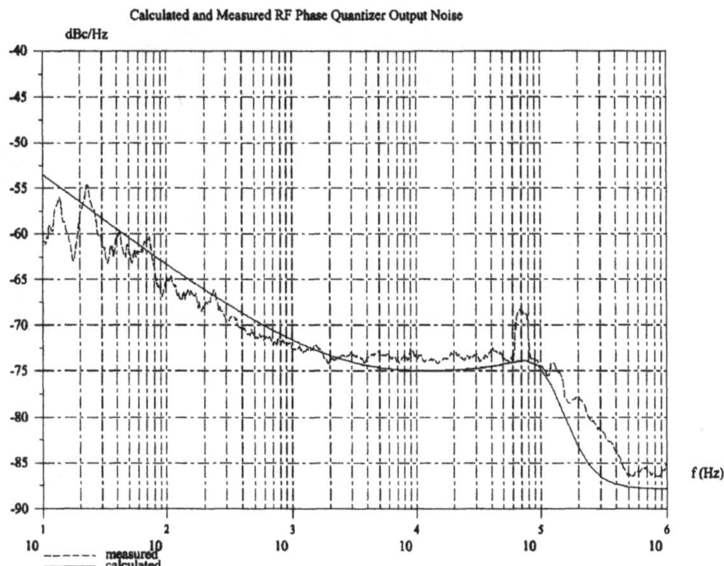

Figure 6.36. Measured RF Phase Quantizer Output Noise

Chapter 7

SUMMARY AND CONCLUSIONS

A new method for the automatic calibration of modulated frequency synthesizers has been developed and presented. The calibration technique operates while the modulator is in service and thus requires no down time for recalibration. Although initially developed for 2 level GFSK, the calibration technique is not specific to this modulation type and it has been experimentally shown to work with 4 level GFSK.

The calibration system was implemented in two forms. The first realization of the system was a board level design. The output center frequency is around 60 MHz with a data rate of 78.125 kbps[1]. The VCO is a discrete transistor design. All digital functions including the RF frequency divider were implemented in a single field programmable gate array (FPGA). The purpose of this implementation was to allow flexibility in the design, the ability to try different approaches, and provide some experimental verification of the simulation results. The second realization was in the form of a full custom BiCMOS integrated circuit. In the integrated circuit implementation, the RF output frequency is around 1.8 GHz and the data rate is 2.5 Mbps for 2 level GFSK and 5.0 Mbps for 4 level GFSK. The integrated circuit includes the calibration circuit and the complete Σ–Δ modulated synthesizer.

The availability of the new calibration architecture allows the Σ–Δ modulated synthesizer to be a viable option for high speed, low power, modulation. By eliminating the need for a factory calibration, the potential manufacturing costs are reduced. In addition, the performance

[1] Using 2 level GFSK. Although not tested on this version of the modulator, twice the data rate should be achievable using 4 level GFSK

over temperature variations is enhanced because the calibration circuit is able to compensate.

7.1. Recommendations for Future Work

Several refinements of the technique presented here are possible. In systems which need to operate in a burst mode where the transmitter is turned on quickly followed by a short burst and then shut off, there may not be time for the calibration circuit to acquire and settle. In this type of system it may be advantageous to implement the gain control error amplifier digitally. It should be possible to periodically run the calibration circuit and then store the resulting setting. When a short burst of modulation is desired, the stored calibration setting is used instead of waiting for the calibration circuit to settle. This approach would provide a tradeoff between calibration accuracy and the time required to bring the modulator online from a standby or sleep mode.

An interesting circuit design problem would be the redesign of the RF phase quantizer circuit for lower power operation. The current design uses a fairly standard bipolar latch which draws static current. However, during most of the time, this current is wasted. In addition, the ECL to CMOS level converter draws static current. Certainly it should be possible to redesign the level converter based around a dynamic sense amplifier and reduce its power consumption to a negligible level. It should also be possible to design the bipolar latch in a way where the static current is shut off for most of the cycle. A power savings of 80 percent should be readily achievable given some time in the design effort.

As identified by [24], the power consumption of the CMOS multimodulus divider is quite large. A redesign of the divider was not a focus of this research and thus remains an open issue. Some preliminary work on a modified design suggests that the power consumption could be reduced by 60 to 80% through the use of CMOS style circuit design techniques in the lower frequency stages of the divider rather than the pseudo-NMOS design which is currently used.

References

[1] Thomas A. D. Riley and Miles A. Copeland, "A simplified continuous phase modulator technique," *IEEE Transactions on Circuits and Systems-II: Analog and Digital Signal Processing*, vol. 41, no. 5, pp. 321–328, May 1994.

[2] Michael H. Perrott, Theodore L. Tewksbury III, and Charles G. Sodini, "A 27–mW CMOS fractional-N synthesizer using digital compensation for 2.5-Mb/s GFSK modulation," *IEEE Journal of Solid-State Circuits*, vol. 32, no. 12, pp. 2048–2060, December 1997.

[3] Thomas H. Lee, *The Design of CMOS Radio Frequency Integrated Circuits.* Cambridge, UK: Cambridge University Press, 1998.

[4] Wayne Struble, Finbarr McGrath, Kevin Harrington, and Pierce Nagle, "Understanding linearity in wireless communication amplifiers," *IEEE Journal of Solid-State Circuits*, vol. 32, no. 9, pp. 1310–1318, September 1997.

[5] Hitoshi Hayashi, Masashi Nakatsugawa, and Masahiro Muraguchi, "Quasi–linear amplification using self phase distortion compensation technique," *IEEE Transactions on Microwave Theory and Techniques*, vol. 43, no. 11, pp. 2557–2564, November 1995.

[6] Hiroshi Yamada, Shiro Ohara, Taisuke Iwai, Yasuhiro Yamaguchi, Kenji Imanishi, and Kazukiyo Joshin, "Self–linearizing technique for L–band HBT power amplifier: Effect of source impedance on phase distortion," *IEEE Transactions on Microwave Theory and Techniques*, vol. 44, no. 12, pp. 2398–2401, December 1996.

[7] Theodore S. Rappaport, *Wireless Communications: Principles and Practice.* Upper Saddle River, New Jersey: Prentice Hall PTR, 1996.

[8] Edward A. Lee and David G. Messerschmitt, *Digital Communication.* Norwell, MA: Kluwer Academic Publishers, 1999.

[9] L. E. Franks, *Signal Theory.* Stroudsburg, Pennsylvania: Dowden and Culver, 1981.

[10] Umberto Mengali and Aldo N. D'Andrea, *Synchronization Techniques for Digital Receivers.* New York, New York: Plenum, 1997.

[11] Kazuaki Murota and Kenkichi Hirade, "GMSK modulation for digital mobile radio telephony," *IEEE Transactions on Communications*, vol. COM-29, no. 7, pp. 1044–1049, July 1981.

[12] J. Klapper, "Demodulator threshold performance and error rates in angle–modulated digital signals," *RCA Review*, vol. 27, no. 2, pp. 226–244, June 1966.

[13] Martin LaCon, "Designing a DECT telephone tests your receiver knowledge," *EDN*, pp. 149–158, October 12, 1995.

[14] Bernard Sklar, *Digital Communications: Fundamentals and Applications.* Englewood Cliffs, New Jersey: Prentice Hall, 1988.

[15] European Telecommunications Standards Institute (ETSI), "Digital enhanced cordless telecommunications (DECT); common interface (CI); part2: Physical layer (PHL)," *ETSI EN 300 175-2 V1.4.2*, 1999.

[16] Philips Semiconductor, "SA900 I/Q transmit modulator data sheet," *RF/Wireless Communications Data Handbook*, 1995.

[17] RF Micro-Devices, "RF2422 direct quadrature modulator data sheet," *1997 Designer's Handbook*, 1997.

[18] Siemens AG, "PMB 2202 mixer DC–2.5 GHz and vector modulator 1.4-2.5 GHz data sheet," *datasheet*, 1995.

[19] Caesar S. Wong, "A 3V GSM baseband transmitter," *1998 Symposium on VLSI Circuits Digest of Technical Papers*, pp. 120–123, 1998.

[20] Alfredo Linz and Alan Hendrickson, "Efficient implementation of an I–Q GMSK modulator," *IEEE Transactions on Circuits and Systems-II: Analog and Digital Signal Processing*, vol. 43, no. 1, pp. 14–23, January 1996.

[21] A. E. Jones, T. A. H. Wilkinson, and J. G. Gardiner, "Effects of modulator deficiencies and amplifier nonlinearities on the phase accuracy of GMSK signalling," *IEEE Proceedings*, vol. 140, no. 2, pp. 157–162, April 1993.

[22] National Semiconductor, "LMX3161 single chip radio transceiver data sheet," *datasheet*, 1996.

[23] Brian Miller and Robert J. Conley, "A multiple modulator fractional divider," *IEEE Transactions on Instrumentation and Measurement*, vol. 40, no. 3, pp. 578–583, June 1991.

[24] Michael H. Perrott, *Techniques for High Data Rate Modulation and Low Power Operation of Fractional-N Frequency Synthesizers.* PhD thesis, Massachusetts Institute of Technology, September 1997.

[25] Conexant, "CX72300-11 spur-free 2.1 GHz dual fractional-N frequency synthesizer," *101217P2*, May 2001.

[26] Hikmet Sari and Said Moridi, "New phase and frequency detectors for carrier recovery in PSK and QAM systems," *IEEE Transactions on Communications*, vol. 36, no. 9, pp. 1035–1043, September 1988.

[27] Robbert Van Der Wal and Leo Montreuil, "QPSK and BPSK demodulator chip-set for satellite applications," *IEEE Transactions on Consumer Electronics*, vol. 41, no. 1, pp. 30–41, February 1995.

[28] Floyd Martin Gardner, *Phaselock Techniques.* New York, New York: John Wiley and Sons, 1979.

[29] Norm Filiol, Neil Birkett, James Cherry, Florinel Balteanu, Christian Gojocaru, Ardeshir Namdar, Tolga Pamir, Kashif Sheikh, Gilles Glandon, Daniel Payer, Ashok Swaminathan, Robert Forbes, Thomas Riley, S. M. Alinoor, Edward Macrobbie, Mark Cloutier, Spyros Pipilos, and Theo Varelas, "A 22 mW bluetooth RF transceiver with direct RF modulation and on-chip IF filtering," *IEEE International Solid-State Circuits Conference: Digest of Technical Papers*, pp. –, 2001.

[30] Anatol I. Zverev, *Handbook of Filter Synthesis.* New York: John Wiley and Sons, 1967.

[31] S. B. Cohn, "Direct-coupled-resonator filters," *Proceedings of the IRE*, vol. 45, pp. 187–196, February 1957.

[32] George L. Matthaei, Leo Young, and E. M. T. Jones, *Microwave Filters, Impedance-Matching Networks, and Coupling Structures.* New York, New York: McGraw–Hill Book Company, Inc., 1964.

[33] Shoji Otaka, Takafumi Yamaji, Ryuichi Fujimoto, Chikau Takahashi, and Hiroshi Tanimoto, "A low local input 1.9 GHz Si-bipolar quadrature modulator with no adjustment," *IEEE Journal of Solid-State Circuits*, vol. 31, no. 1, pp. 30–37, January 1996.

[34] C. Vaucher and D. Kasperkovitz, "A wide-band tuning system for fully integrated satellite receivers," *IEEE Journal of Solid-State Circuits*, vol. 33, no. 7, pp. 987–997, July 1998.

[35] D. Hitko, T. Tewksbury, and C. Sodini, "A 1 V, 5 mW, 1.8 GHz, balanced voltage–controlled oscillator with an integrated resonator," *International Symposium on Low Power Electronics and Design Proceedings*, pp. 46–51, August 1997.

[36] Wolfgang Heimsch, Birgit Hoffmann, Roland Krebs, Ernst G. Müllner, Bruno Pfäffel, and Klaus Ziemann, "Merged CMOS/bipolar current switch logic (MCSL)," *IEEE Journal of Solid-State Circuits*, vol. 24, no. 5, pp. 1307–1311, 1989.

[37] Hooman Reyhani and Philip Quinlan, "A 5 V, 6-b, 80 Ms/s BiCMOS flash ADC," *IEEE Journal of Solid-State Circuits*, vol. 29, no. 8, pp. 873–878, 1994.

[38] Muhammad S. Elrabaa and Mohamed I. Elmasary, "Design and optimization of buffer chains and logic circuits in a BiCMOS environment," *IEEE Journal of Solid-State Circuits*, vol. 27, no. 5, pp. 792–801, 1992.

[39] S. H. K. Embabi, A. Bellaouar, and M. I. Elmasary, "Analysis and optimization of BiCMOS digital circuit structures," *IEEE Journal of Solid-State Circuits*, vol. 26, no. 4, pp. 676–679, 1991.

[40] Peter J. Lim and Bruce A. Wooley, "An 8–bit 200–MHz BiCMOS comparator," *IEEE Journal of Solid-State Circuits*, vol. 25, no. 1, pp. 192–199, February 1990.

[41] Bendik Kleveland, Thomas H. Lee, and S. Simon Wong, "50-GHz interconnect design in standard silicon technology," in *IEEE MTT-S International Microwave Symposium Digest*, June 1998.

[42] Bernard Sklar and fred harris, "Advanced communications systems using digital signal processing." UCLA Course Notes, Engineering 881.123, July 1995.

[43] Ronald E. Crochiere and Lawrence R. Rabiner, *Multirate Digital Signal Processing.* Englewood Cliffs, New Jersey: Prentice Hall, 1983.

[44] Eugene B. Hogenauer, "An economical class of digital filters for decimation and interpolation," *IEEE Transactions on Acoustics, Speech, and Signal Processing*, vol. ASSP-29, no. 2, pp. 155–162, April 1981.

[45] Alan Oppenheim and Ronald Schafer, *Discrete-Time Signal Processing.* Englewood Cliffs, New Jersey: Prentice Hall, 1989.

[46] Chris Dick and Fred Harris, "FPGA DSPs–the platform for NG wireless communications," *RF Design*, pp. 56–66, October 2000.

[47] Anantha P. Chandrakasan, Samuel Sheng, and Robert W. Brodersen, "Low-power CMOS digital design," *IEEE Journal of Solid-State Circuits*, vol. 27, no. 4, pp. 473–484, April 1992.

[48] Kai Hwang, *Computer Arithmetic – Principles, Architecture, and Design.* New York, New York: John Wiley and Sons, 1979.

[49] Henry Samueli, "An improved search algorithm for the design of multiplierless FIR filters with powers–of–two coefficients," *IEEE Transactions on Circuits and Systems*, vol. 36, no. 7, pp. 1044–1047, July 1989.

[50] Richard I. Hartley, "Subexpression sharing in filters using canonic signed digit multipliers," *IEEE Transactions on Circuits and Systems-II: Analog and Digital Signal Processing*, vol. 43, no. 10, pp. 677–688, 1996.

[51] John Van de Vegte, *Feedback Control Systems.* Englewood Cliffs, NJ: Prentice Hall, 1990.

[52] Kuniharu Uchimura, Toshio Hayashi, Tadakatsu Kimura, and Atsushi Iwata, "Oversampling A-to-D and D-to-A converters with multistage noise shaping modulators," *IEEE Transactions on Acoustics, Speech, and Signal Processing*, vol. ASSP-36, pp. 1899–1905, December 1988.

[53] Frederick Emmons Terman, *Electronic and Radio Engineering.* New York: McGraw-Hill Book Company, 1955.

[54] Daniel R. McMahill, *Automatic Calibration of Modulated Fractional-N Frequency Synthesizers.* PhD thesis, Massachusetts Institute of Technology, June 2001.

[55] Daniel R. McMahill and Charles G. Sodini, "Automatic calibration of modulated $\Sigma - \Delta$ frequency synthesizers," *2001 Symposium on VLSI Circuits Digest of Technical Papers*, pp. 51–54, 2001.

[56] Daniel R. McMahill and Charles G. Sodini, "A 2.5 Mbps GFSK, 5.0 Mbps 4-FSK automatically calibrated Σ–Δ frequency synthesizer," *IEEE Journal of Solid-State Circuits*, to be published.

Appendix A
Tools

This appendix gives a brief overview of some of the computational tools used during the design, testing, and typesetting of this work. The software used was a mix of commercial software and free software. All of the free software was available via the NetBSD Packages Collection.[1]

A.1. Simulation Tools

The large range of time constants present in the calibration system present a challenge in the numerical simulation of the system. The RF output is nearly 2 GHz, the phase/frequency detector and charge pump operate at 20 MHz, the main synthesizer loop bandwidth is around 80 kHz, the digital phase locked loop has a loop bandwidth of around 1 kHz, and the calibration loop bandwidth is a few Hertz. This extremely large range of time constants (9 decades!) makes a spice simulation impossible. The complexity of the integrated circuit, around 15,000 devices, makes some sort of computational verification an absolute requirement.

The initial investigations into the automatic calibration approach were simulated using *Matlab*.[2] To obtain the gain error detector output versus gain error curves, the ideal modulation phase trajectory and idealized RF phase quantizer output were generated from equivalent baseband models. The track/hold circuit in the gain error detector can not be modeled as a vectorized operation in *Matlab*[3] so a C-mex function was used for that portion. These simulations omitted the phase tracking (digital phase locked loop) portion of the gain error detector.

The next level of modeling included the DPLL operation, Σ–Δ modulator, and main synthesizer dynamics. Due to the nonlinear[4] feedback nature of the system, Matlab was no longer suited for the simulation. A custom simulation engine written by Mike Perrott in the C programming language was used as the starting point towards

[1] `http://www.netbsd.org/Documentation/software/packages.html`

[2] `http://www.mathworks.com`

[3] Vectorized operations are essential in *Matlab* for reasonable processing speed

[4] The phase/frequency detector in the main synthesizer and in the DPLL are non-linear and have memory associated with them.

writing a custom simulator. This simulator modeled the Σ–Δ modulator, DPLL, and other digital signal processing circuits at the bit level and the analog blocks at a behavioral level. The program *graph* which is part of the GNU plotutils package[5] was used for simple graphical output. *Scilab*-2.5[6] was used for more sophisticated post processing and plotting of the simulator output.

The symbolic algebra program *yacas*[7] was used to check several of the analytic calculations.

During the circuit design phase of the project, extensive use of a spice based circuit simulation program was used. The simulation of individual blocks with this type of simulator is feasible. The circuit simulator used includes the ability to perform switch level simulation of the CMOS logic circuits. Various inputs to the digital signal processing sections were generated using the freely available Icarus Verilog[8] compiler. The test stimulus was used to drive both the switch level circuit simulation[9] and a higher level verilog simulation of the system. The two outputs were then compared to verify correct logical operation of the circuit.

The physical layout of the chip was done using layout software from Cadence.[10] Design rule checks and layout versus schematic checks were performed using *DRACULA*, also from Cadence.

A.2. Test Setup Tools

The schematic capture and printed circuit board layout tools from Accel EDA[11] were used for the physical design of the test board. The FPGA design was done in verilog and *Synplify*[12] was used to synthesize a Xilinx netlist. The Xilinx netlist provides the input to the place and route tools from Xilinx which generate a prom file. The microprocessor code development was done in C and compiled with the Bytecraft C compiler.[13]

The control software which provides the user interface to the test board was written in C and runs under the freely available NetBSD[14] operating system on a SparcClassic which was rescued from the trash.

The varactor measurements were automated using *IC-CAP*. The program *fasthenry*[15] was used to assist in extracting the interconnect inductance.

A.3. Typesetting Tools

The final typesetting of this document was done using LaTeXrunning on a 500 MHz PC164 Alpha workstation. The operating system is NetBSD-1.5.[16] The schematic

---

[5] `http://www.gnu.org/software/plotutils/plotutils.html`

[6] `http://www-rocq.inria.fr/scilab/`

[7] `http://www.xs4all.nl/~apinkus/yacas.html`

[8] `http://icarus.com/eda/verilog/index.html`

[9] The switch level netlist is generated from the master schematics which also are used for layout versus schematic checking. This insures that the simulated circuit matches the layout.

[10] `http://www.cadence.com`

[11] `http://www.acceltech.com`

[12] `http://www.synplicity.com`

[13] `http://www.bytecraft.com`

[14] `http://www.netbsd.org`

[15] `http://kontiki.mit.edu/rle/research/info_research_proj.html`

[16] `http://www.netbsd.org`

and block diagrams were drawn using *Tgif*-4.1.40[17]. The photographs of the test boards were shot on color 35mm film and scanned using the *xsane*[18] plug-in to *The GIMP*.[19] *The GIMP* was then used to cut away the background from the images. The HPGL format plots captured from the spectrum analyzer, oscilloscope, and vector modulation analyzer were converted to postscript using the program *hp2xx*.[20]

A.4. Miscellaneous

Most of the simulations, results post processing, and plotting were run by *make*. In addition, standard Unix tools such as *awk*, *sed*, and *sh* were used to help automate the conversion of test data and interface various simulation tools. The availability of a computing environment which supports programmable command line utilities such as these was a requirement for efficient simulation of this system. Hopefully the trend towards monolithic GUI-only oriented operating systems and simulation tools will not cause the demise of this powerful flexibility.

[17] `http://bourbon.cs.umd.edu:8001/tgif/tgif.html`
[18] `http://www.wolfsburg.de/~rauch/sane/sane-xsane.html`
[19] `http://www.gimp.org`
[20] `http://www.gnu.org/software/hp2xx/hp2xx.html`

Index

Zeitfracht Medien GmbH
Ferdinand-Jühlke-Straße 7
99095 Erfurt, Deutschland
produktsicherheit@kolibri360.de